Pro Tools 9

Music Production, Recording, Editing, and Mixing

Mike Collins

D1342139

ELSEVIER

AMSTERDAM • BOSTON • HEIDELBERG • LONDON • NEW YORK • OXFORD • PARIS
SAN DIEGO • SAN FRANCISCO • SINGAPORE • SYDNEY • TOKYO

Focal Press is an imprint of Elsevier

Focal Press

Focal Press is an imprint of Elsevier
225 Wyman Street, Waltham, MA 02451, USA
The Boulevard, Langford Lane, Kidlington, Oxford, OX5 1GB, UK

Notices

Knowledge and best practice in this field are constantly changing. As new research and experience broaden our understanding, changes in research methods, professional practices, or medical treatment may become necessary.

Practitioners and researchers must always rely on their own experience and knowledge in evaluating and using any information, methods, compounds, or experiments described herein. In using such information or methods they should be mindful of their own safety and the safety of others, including parties for whom they have a professional responsibility.

To the fullest extent of the law, neither the Publisher nor the authors, contributors, or editors, assume any liability for any injury and/or damage to persons or property as a matter of products liability, negligence or otherwise, or from any use or operation of any methods, products, instructions, or ideas contained in the material herein.

Library of Congress Cataloging-in-Publication Data
Collins, Mike, 1949–
    Pro Tools 9: music production, recording, editing, and mixing/Mike Collins.
        p. cm.
    ISBN 978-0-240-52248-7
1. Pro Tools. 2. Digital audio editors. I. Title.
    ML74.4.P76C66 2012
    781.3'4536—dc22   2011008686

British Library Cataloguing-in-Publication Data
A catalogue record for this book is available from the British Library.

ISBN: 978-0-240-52248-7

For information on all Focal Press publications
visit our website at *www.elsevierdirect.com*

Printed in the United States of America

11 12 13 14 15   5 4 3 2 1

Typeset by: diacriTech, Chennai, India

Working together to grow
libraries in developing countries

www.elsevier.com | www.bookaid.org | www.sabre.org

ELSEVIER   BOOK AID International   Sabre Foundation

Contents

Contents

Contents

Contents

Contents

Contents

About the Author

Mike Collins is a studio musician, recording engineer, and producer who has worked with all the major audio and music software applications on professional music recording, TV, and film scoring sessions since 1988. During that time, Mike has regularly reviewed music and audio software and hardware for magazines, including PRS for *Music's M* magazine, *Future Music, Computer Music, Macworld, MacUser, Personal Computer World, Sound On Sound, AudioMedia, Studio Sound, Electronic Musician, EQ, MIX,* and others. Mike also writes industry news and technical reports for Pro Sound News Europe.

Mike has been writing for Focal Press since 2000. His first book *Pro Tools 5.1 for Music Production,* was published in December of that year. His second book *A Professional Guide to Audio Plug-ins and Virtual Instruments* was published in May 2003. *Choosing & Using Audio & Music Software,* Mike's third title for Focal Press was published in 2004 along with *Pro Tools for Music Production Second Edition.* Mike's fifth book *Pro Tools LE & M-Powered* was published in the summer of 2006 while *Pro Tools 8* was published in the summer of 2009.

In the second quarter of 2010, Mike was invited to join a team of audio transfer engineers at Iron Mountain's Xepa Digital Studios in Slough to help to transfer a large part of Universal Music's back catalogue of classic popular recordings from analogue and digital tape copies of the archive stored in the UK to WAV files for archiving at Iron Mountain's secure underground facility in the US. Using an Ampex ATR100 tape machine, Mike personally transferred from ¼-inch tape much of the Chess catalogue along with many albums by Cat Stevens, Joe Jackson, Louis Armstrong, Quincy Jones, Barry White, Tricia Yearwood, and lots of other famous bands and artists from Universal's library of "hit" recordings from the past 60 years.

Since the summer of 2010, Mike has been writing songs, recording, and performing "live" with vocalist Aurore Colson as "Michael & Aurora"—see www .michaelandaurora.com for more info.

Also active as a Music Technology Consultant, Mike Collins regularly presents seminars and chairs discussion panels on Pro Tools, music production, music technology, music rights, and copyrights.

Contact Details

The author may be contacted via email at mike@mikecollinsmusic.com. The author's website can be found at www.mikecollinsmusic.com and a professional profile is available at www.linkedin.com/in/mikecollinsmusic.

Acknowledgments

First of all, I would like to thank Catharine Steers at Focal Press for commissioning this book. I am also grateful for the efforts of all at Focal Press who are involved with publishing and marketing my books.

Thanks also to Avid Pro Audio Application Specialist Simon Sherbourne, who provided helpful clarification of technical issues, and to Tim Hurrell, Charlotte Dawson, Simon Caton, and Sam Butler who all "went the extra mile" to supply software and technical information over the Christmas and New Year holiday period during the preparation of this book.

Special thanks to Matt Ward, President at Universal Audio and Erik Hanson, Director of Marketing at Universal Audio for supplying a UAD-2 system with its suite of plug-ins for Pro Tools for me to evaluate alongside Pro Tools 9 while writing this book.

I would like to thank John Leckie for pointing out a couple of things that I had overlooked in the Mixing chapter—explaining some basic points about mixing consoles that he had been shown during his training at Abbey Road Studios that really helped to clarify everything for me. Thanks also to Jean-Paul "Bluey" Maunick, Kirk Whalum, and Dario Marianelli who took time out from their busy schedules to review and comment on the original edition of this book prior to publication. I would especially like to thank the technical reviewer, Lorne Bregitzer, who not only picked up on several important errors and omissions, but also put forward several small, but useful points to add into the text.

I would also like to thank all the members of my family, most importantly my father Luke Collins, my mother Patricia Collins, my brothers Anthony, and Gerard Collins, and all my friends, in particular Barry Stoller, Keith O'Connell, Sia Duma, Anthony Washington, and Tom Webster.

Last, but not least, a very special thanks must go to my "muse," Aurore Colson, for all her tremendous support and encouragement throughout the time that I was preparing for and then completing this book.

Mike Collins, January 2011

In This Chapter

Pro Tools—The World's Leading Digital Audio Workstation

Introduction

Pro Tools digital audio workstation systems are used all over the world in professional recording situations for "live" recording at gigs and concerts, for film scoring, for audio-for-video, and for audio postproduction, and so on.

This book will be particularly useful for people who are upgrading from earlier versions of Pro Tools or "cross-grading" from other digital audio workstations. It will also serve as a useful handbook to use alongside any Pro Tools system to clarify how things work and to provide useful operational tips and notes.

Although this book is not intended to be an entry-level text, I have explained concepts as simply as I can wherever possible. Mindful of the fact that some readers may not have English as their first language, I have tried to avoid colloquialisms and slang as much as possible and have done my best to write as clearly as possible. I hope that more experienced audio professionals will regard the sometimes lengthy explanations of what they may consider to be basic points as useful reminders about how things work so that they can skip past easily enough if they wish.

There are several excellent books already available that may be more suitable for beginners, including my own *Pro Tools LE & M-Powered,* which takes a more step-by-step approach. And the Pro Tools Reference Guide and other Help documents are always available from the Pro Tools Help menu.

This book is about *music production* using Pro Tools, which is where Avid's focus for Pro Tools development has shifted since the release of Pro Tools version 8. This is not to suggest for a moment that Pro Tools is not the leading system used for adding sound and music to picture or for mixing in surround. It is just a reflection of the fact that Pro Tools has truly come of age as a music production system, and the greater part of my experience is with music production, so I am naturally inclined to write about the things that I know most about.

Despite the fact that surround formats have been with us for many years now, the main focus of the book is firmly on using Pro Tools for music production and mixing in stereo. The reality is that DVD-Audio, Super Audio CD (SACD), and other formats that can deliver surround sound are still "niche" products. When did you ever hear of a Top 40 "hit" being released primarily as a 5.1 surround mix and making it because of this? I never heard of this happening—not one time! So, this book will not digress into the fascinating world of surround sound—as interesting as this is to me and to many people I know.

Having read all of the Pro Tools manuals and documentation in depth, along with most of the other books written about Pro Tools, I have noticed that some areas are not covered as thoroughly as or as accurately as they could be, so I have made a point of addressing as many of these areas as possible. One example would be levels and metering, which are covered in detail in chapter 7.

You will not find every feature of Pro Tools covered in this book: it makes no attempt to be "all things to all people." What you will find are clear explanations of most of the things you definitely need to know about to record, edit, and mix audio and MIDI in Pro Tools in stereo. There is a substantial chapter on MIDI that covers all the technical stuff about Pro Tools that you need to get professional results, whether using hardware or software MIDI instruments.

Wherever possible, I have tried to explain topics that are covered in the Reference Guide in a clearer or alternative way. Throughout the book you will find highlighted Tips and Notes. The tips are often taken from my personal experience, and the notes are often technical points taken from the Pro Tools Reference Guide or other documentation that may otherwise be overlooked, so watch out for these.

Learning Pro Tools

The only way to learn Pro Tools is to use Pro Tools as often as possible in as many contexts as possible. Even if you own a Pro Tools HD system, I strongly recommend that you get a laptop with Pro Tools 9 software, so you can take it wherever you go. This way you can practice recording, editing, sequencing, and mixing Pro Tools sessions, or making music with virtual instruments and loops or whatever, while you are away from your main studio setup. Or you could get a compact desktop such as an iMac at home and practice using this with Pro Tools 9 and an Mbox or other compact hardware.

If you are really determined to be the fastest Pro Tools operator around, you will have to "eat, sleep, and breathe" Pro Tools until you have learned as many of the keyboard commands as you can remember and get as much varied experience of recording as possible.

On the other hand, you may prefer to take a more relaxed approach—exploring the menus at a more leisurely pace rather than constantly typing commands on the keyboard. And Pro Tools has the advantage that it is simpler to get familiar with than most of its competitors—especially as it has just two main screens, the Mix and Edit windows.

Pro Tools 9

In November 2010, Avid took the Pro Tools community completely by surprise by unveiling the new Pro Tools|HD Native hardware that uses the much more powerful processors in today's desktops and laptops for digital signal processing instead of the dedicated DSP chips used on the Pro Tools|HD cards.

No longer are Pro Tools|HD users compelled to buy the more expensive Pro Tools|HD cards with their dedicated DSP processors—unless they choose to do so and can afford to do so! Now, Pro Tools users can put Pro Tools|HD Native systems together using the much more affordable PCIe card along with powerful desktop computers (such as a top-of-the-range Mac Pro with a pair of 2.93 GHz 6-core Intel Xeon "Westmere" processors) to handle all the processing. And, even better, Pro Tool|HD Native works with Mac or Windows, with Pro Tools HD software, or with third-party software. So, Pro Tools|HD Native owners can now use third-party software as they please—even Logic Studio will run on this card!

Figure 1.1 Pro Tools 9 "splash" screen.

Then, just as we were getting over the shock of this breakthrough product, the "revolutionary" new Pro Tools 9 software was announced later that month! OK, here's the revolutionary bit: Pro Tools 9 can now run stand-alone using the audio hardware built into your computer, or with third-party audio interfaces, or with Avid hardware such as Pro Tools|HD Native. But you are not compelled to use Avid hardware any more—as was the case with Pro Tools HD and LE—you can even run Pro Tools 9 on your laptop's built-in soundcard!

Even more importantly for many users, Pro Tools 9 now has most of the advanced features of Avid's "flagship" Pro Tools HD software, including automatic delay compensation, much higher track count, EuCon support, etc.

So, Pro Tools 9 will let you create bigger, better-sounding mixes with more tracks and extend your workflow to projects created in other audio and video software. It will also work with all Avid (formerly Euphonix) Artist Series, Pro Series mixing consoles, and DAW controllers, which is a "big deal" for many professional and project studio users who have discovered the benefits of these consoles and controllers.

And there are lots of smaller, yet significant improvements. With earlier versions of Pro Tools, you had to pay extra to get the MP3 Export Option and to get OMF and AAF Import and Export Options. These features are now included both in Pro Tools HD 9 and in Pro Tools 9, making it much easier, for example, to use a laptop with Pro Tools 9 for sound effects "spotting." Simply import an AAF that was exported from Avid Media Composer along with a Quicktime file containing the picture and you can get started right away without having to buy the expensive DigiTranslator option that was required earlier.

Also, the AAF features have been significantly improved by adding the ability to import stereo AAF audio tracks and to import RTAS plug-in data from AAF sequences—for example, from Avid Media Composer V. 5 or higher sessions. This makes it much easier for a video post editor to get things started with an audio mix in Media Composer and then get this over to the audio post editor working in Pro Tools, not only with the track volumes and pan settings but also with the RTAS plug-ins all set up as they were in Media Composer. There is also a new "Locators To Import" setting for importing locators from Media Composer-generated AAF sequences. These appear as Markers in Pro Tools which further helps to improve this workflow.

Pro Tools 9 offers a choice of several pan depths that you can select in the Session Setup window. The default setting is −3 dB (it was −2.5 dB in earlier versions), and this can be adjusted from −2.5 to −6 dB to provide compatibility with most other systems.

Autoscrolling for tracks in the Mix and Edit windows is now provided for Pro Tools 9, along with the advanced Beat Detective features including separate multiple tracks and Collection mode, that were previously available only in Pro Tools HD.

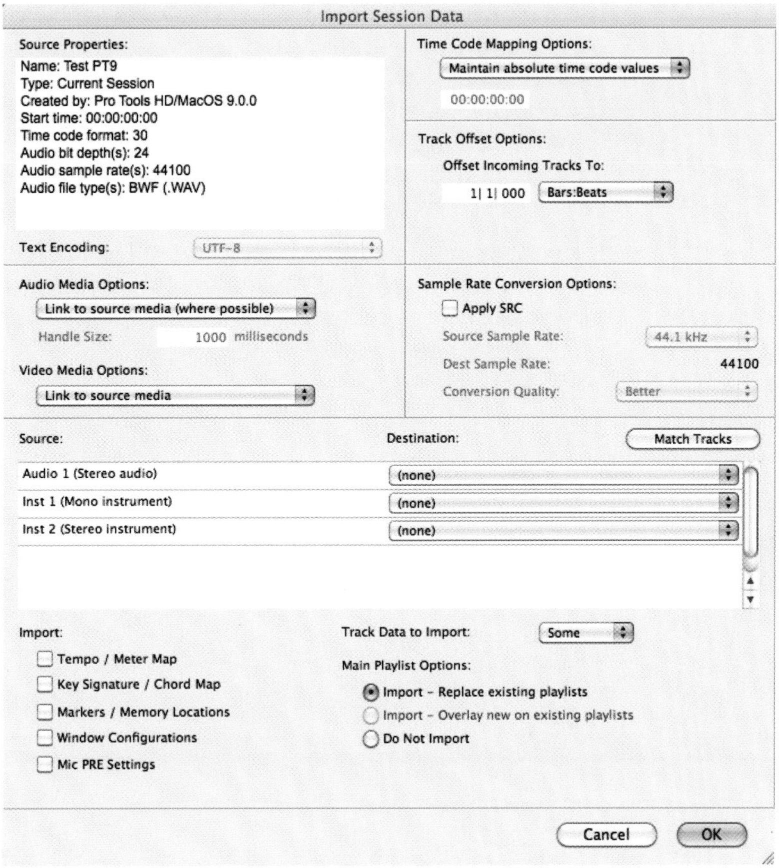

Figure 1.2
Import Session
Data dialog.

Pro Tools 9 also has the Import Session Data feature that is so useful for bringing data from other Pro Tools sessions into your current session. Earlier this was only available in Pro Tools HD or in the expensive Toolkits for LE.

What's New in Pro Tools 9 for Pro Tools LE Users?

Compared with Pro Tools LE, Pro Tools 9 has double the number of voice-able audio tracks (96 mono or 48 stereo tracks) and allows you to have up to 128 Audio Tracks in a session, although only the first 96 mono or 48 stereo (or a mixture of these) will be operational at the same time.

PT 9 also has eight times the number of internal busses (256), double the number of instrument tracks (64), and can support up to 160 auxiliary tracks and doubles the number of MIDI tracks to 512.

Pro Tools 9 now has support for PRE, Avid's remote-controllable 8-channel mic preamp and has the advanced DigiBase search features and Catalogs, Autoscrolling for tracks in the Mix and Edit windows, advanced Beat Detective features including Collection mode, the Delay Compensation features for Auxiliary Input tracks and for MIDI, and the "Hardware Insert Delay Compensation" and "Low Latency Monitoring During Recording" features that were missing from Pro Tools LE compared with Pro Tools HD.

For audio-for-video postproduction, Pro Tools 9 now has all the advanced features that earlier were restricted to Pro Tools|HD or the LE Toolkits, including Timebase Rulers, Time Code Rulers, Feet+Frames, and Pull Up and Pull Down commands for both audio and video.

What's New in Pro Tools 9 for Pro Tools HD Users?

For LE users it's almost a "no-brainer" that they should upgrade. But what about Pro Tools HD users? What's in it for these guys?

Well, for a start, the new features make it much easier to collaborate with other audio and video software users because you can now run third-party software on Pro Tools|HD Native hardware or run Pro Tools 9 software on third-party hardware.

For high-end film and video work, Pro Tools HD 9 (and Pro Tools 9 with Complete Production Toolkit 2) now features 7.1 and 7.0 standard HD surround formats. The new variable stereo pan depths feature improves mixing in the stereo field, and a new DTS-style 7.1 surround panner with true side pan will enable users to create better surround mixes.

Version 9 also offers new Track and Send Output selector commands for faster, simplified signal routing and mixer configuration, and an increased number of audio, aux, and MIDI tracks and internal mix busses.

Most importantly for many users, Pro Tools HD 9 has double the number of audio tracks for a total of 512, double the number of MIDI tracks for a total of 512, and double the number of internal busses for a total of 256, along with up to 128 instrument tracks and 160 auxiliary tracks.

Also, professional users will definitely appreciate the hands-on mixing options now available using the Avid Artist Series or Pro Series (formerly Euphonix) control surfaces. Avid bought Euphonix in April 2010 and added support for EuCon with Pro Tools 9. Pro Tools 9 supports EuCon natively to allow EuCon commands to directly connect to the application for full control integration.

Figure 1.3 Artist Series Controllers: MC Transport, MC Control, and up to four MC Mix units can snap together to build larger mixing and editing systems in any combination.

For more info about Euphonix, see Appendix 3 on the accompanying website for this book or explore Avid's Euphonix websites at www.euphonix.com/pro/ and www.euphonix.com/artist/.

Should Existing Users Upgrade?

If you have Pro Tools LE or M-Powered but you don't already have all the extra toolkits, there can only be one answer to this question—and it's yes! Being able to use most of the features that were earlier reserved for Pro Tools HD users totally justifies this upgrade—in my view.

Pro Tools 9 costs £504.95/$599, with "street" prices as low as £380 in the UK. It replaces Pro Tools LE software, which will no longer be developed, although LE will continue to be supported for sometime to come with customer service updates. Crossgrades to PT 9 for LE users will cost around £209.95/$249, with "street" prices as low as £180 in the UK.

If you want to be able to work on full-blown HD sessions, playing back up to 192 tracks anywhere, anyplace, anytime, then you will need the Complete Production Toolkit 2, which costs £1430 + VAT in the UK or $1995 US. This price is high enough to deter many users, but professional users who want to put large sessions on a laptop so that they can "tweak" these on a plane, for example, will certainly regard this as a very desirable purchase.

The upgrade price for Pro Tools HD users to go to Pro Tools HD 9 is also very reasonable at $349/£293.95, with "street" prices of about £250 in the UK. Pro Tools HD 9 software does offer Pro Tools HD users access to new features, as well as the ability to run the HD software stand-alone—without the Pro Tools HD hardware—although some feature limitations apply when not using the HD hardware (e.g., no TDM plug-ins).

> ## Note
>
> Be aware that the Pro Tools|HD system iLok must be attached to the computer in order to run the Pro Tools HD 9 software stand-alone, for example, if you want to take this out and about on your laptop. Also, the Pro Tools HD 9 software upgrade requires your Pro Tools|HD Core card serial number and your Pro Tools HD 7 or 8 iLok license. The Pro Tools HD 9 upgrade license requires surrender of your earlier Pro Tools HD iLok license. The new license replaces the earlier license with a single bundled license that contains authorizations for Pro Tools HD 9 and Pro Tools HD 8 (and Pro Tools 9 with Complete Production Toolkit 2 features when using your license on a system without Pro Tools|HD or Pro Tools|HD Native hardware).

Summary

Avid has rewritten the rules for Pro Tools owners in ways that are certain to broaden the user base significantly.

More users can buy Avid hardware such as Pro Tool|HD Native, safe in the knowledge that they can use whatever software they like with this—and more users can buy Pro Tools 9, safe in the knowledge that this will work with just about any hardware they are likely to have available!

For existing users, the version 9 software includes several useful improvements, while keeping the user interface that was developed for Pro Tools 8 intact. This user-interface was not in any way "broken"—so, there was definitely no need to fix it!

Also, being able to take large Pro Tools|HD studio sessions away and open these on a laptop using Pro Tools 9 with the Complete Production Toolkit 2 really does open the door to much improved workflows with new levels of convenience for users.

All in all, I get the sense that Avid is breathing "new life" into Pro Tools with this more "open" approach, which bodes well for the future.

For more info about Pro Tools and related products, see Avid's website at www.avid.com/US/resources/digi-orientation.

In This Chapter

Using Pro Tools

Introduction

Getting started with any software environment involves something of a learning curve and, if you are new to Pro Tools, you will find that this is no exception. Even if you have just upgraded from an earlier version, there may be a number of new features that you have not encountered earlier. And if you are "cross-grading" from another platform such as Apple Logic, MOTU Digital Performer, or Steinberg Cubase, you may be looking for features or trying to use keyboard commands that you are used to and wondering where these are in Pro Tools. So, boot up your Pro Tools system and spend as much time as possible working "hands-on" with the software until you know it inside out! Oh, and reading this book will help.

Getting Started

If you are using Pro Tools for the first time, you will need to spend some time familiarizing yourself with the user interface, or with its more recent features if you are upgrading. You will also probably find it helpful to configure the software to suit the ways you like to work. You should take some time out to practice using the system in a noncritical situation first (without impatient clients breathing down your neck) and take the trouble to learn at least a basic set of keyboard commands so that you don't have to use the mouse and menus all the time. You should also learn how to quickly find your way around whichever piece of music you are working on—zooming in and out and navigating along the timeline until you feel comfortable with all this.

Help

Avid provides excellent manuals for Pro Tools as Adobe Acrobat .pdf document files that you can access from the Help menu. When you open the Pro Tools Reference Manual, for example, the Acrobat application launches, and the document file opens into a new window on your computer. You should resize this window and position it in a convenient place on your screen.

It can be very useful to keep the relevant manual open on your computer, but hidden until you need it. You can use the standard menu command or the keyboard Command-H on the Mac to hide the Acrobat application. When you want to reveal the manual again, you can always select it again from the Help menu or use the Show All menu command. One of the quickest ways to do this is to press and hold the Command key, and then repeatedly press the Tab key until you see the application you want to be displayed in the middle of the screen. When the Tab has moved you along the list of open applications to Acrobat, let go of the Command key, and Acrobat will be shown and brought to the front. Windows users will have their own preferred ways of doing these things.

Starting the Engines

As with any complex piece of machinery, there are various settings that you may need to adjust each time you want to use it. This is certainly the case with Pro Tools. I recommend that you always check the Playback Engine settings and the I/O Setup at the start of any new session. You may also need to make some changes to the Hardware Setup if you want to hook up additional hardware.

The Playback Engine

Most of the settings in the Playback Engine dialog default to sensible choices, but you do need to become familiar with the settings that you can make here if you want to get the best out of your Pro Tools system.

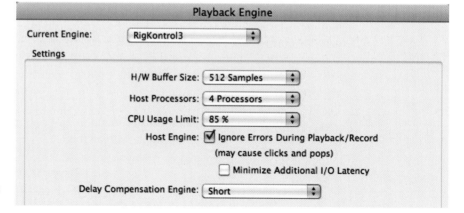

Figure 2.1 Part of Playback Engine dialog for Pro Tools 9 showing the Current Engine pop-up.

The Hardware Buffer Size setting controls the size of the hardware memory buffer used for host-based tasks such as RTAS plug-in processing. If you are recording "live" input, or if you are playing or recording using RTAS virtual instruments, you should choose a lower Hardware Buffer Size, such as 128 samples, which reduces monitoring latency to its minimum. When you are mixing, and especially if you are using lots of RTAS plug-ins, you should choose a higher size, such as 1024 or 2048 samples.

> **Note**
>
> Higher Hardware Buffer sizes can affect the accuracy of plug-in automation, mute data, and the timing for MIDI tracks, especially on slower computers.

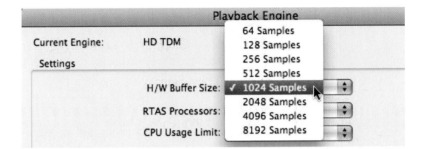

Figure 2.2 Pro Tools HD Playback Engine Hardware Buffer Size.

If your computer has more than one processor, you can choose the Host/RTAS processing that uses one or more of these. You can also limit the percentage of CPU resources allocated to Pro Tools host processing tasks. Higher settings allocate more processing power to Pro Tools so you can work with larger sessions or use more real-time plug-ins. Lower settings allocate more processing power to things such as screen redraws and are particularly useful when you are running other applications at the same time as Pro Tools.

With Pro Tools|HD systems, one of the most important settings to make here is the number of voices, and how these are allocated to the DSPs in your system. The number of voices that you choose here controls the number of tracks that can play back simultaneously.

If you don't need too many tracks, you should choose a lower number of voices and DSPs. With larger sessions, especially with expanded systems, you can use a higher number of voices.

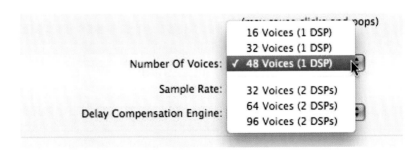

Figure 2.3 Pro Tools HD Playback Engine dialog: setting the Number Of Voices.

You can also adjust the DAE Playback Buffer Size to tweak the disk buffers and to optimize system performance while recording or playing back. Lower settings improve recording and playback initiation speed, but can make it difficult for slower hard drives to record or play reliably. Higher settings allow for higher densities of edits during a session, but the time lag between pressing play/record and playback or recording actually commencing can become noticeable. A time lag can also occur when you are editing during playback, so you do have to be careful with this setting.

The Delay Compensation Engine can be used to compensate for the latency delays introduced by plug-ins. Depending on the way you are using your plug-ins, you can either turn this off to preserve CPU power or use the shorter or longer delay compensation settings. For example, it makes sense to keep this off when you are tracking 'live' audio, as the Delay Compensation will introduce a delay in monitoring.

Automatic Delay Compensation

With the Delay Compensation Engine enabled, Pro Tools automatically compensates for any delays between tracks introduced by plug-ins or routings.

Phase problems have been an issue for Pro Tools users in the past, especially when using plug-ins on some tracks and not on others, or when using different plug-ins with different latency delays, thus introducing phase shifts as a result of the short delays between these tracks. This could be particularly problematic on drum kits that have been recorded using more than one microphone if, for example, you then put a compressor on the snare and not on the rest of the kit, consequently introducing a delay on the snare relative to the rest of the kit. Similarly, if complex routings were used with some tracks and not with others, this would cause delays between these tracks due to the additional time taken to route signals through the busses or inserts.

Another example of problematic delays would occur if you record a bass guitar to separate channels using a DI box for one channel and a microphone in front of a loudspeaker for the other channel. If you then insert a look-ahead peak limiter plug-in (such as Avid's Maxim) on one of these channels, this will delay the output of this track (by 1024 samples with Maxim) compared with the other track, causing phase cancellations when you mix these together.

Fortunately, all Pro Tools systems feature Automatic Delay Compensation. When you activate this feature, all your audio will remain time aligned and in phase, no matter what routings or plug-ins are used.

It is (and was always) possible to compensate for these delays by delaying all the undelayed tracks by the same amounts as the delayed tracks. The problem is that this takes time to do, and it is all too easy to overlook such details on a busy session.

Now, you have the option of enabling the Automatic Delay Compensation feature in the Playback Engine dialog. This automatically makes adjustments in Pro Tools HD software to compensate for any latencies in the I/O, internal and external routing, and plug-in algorithm processing, ensuring that recordings and mixes stay perfectly time aligned and phase accurate.

Figure 2.4
Enabling
Automatic Delay
Compensation
in the Playback
Engine dialog.

Manual Delay Compensation

TDM plug-ins typically have small processing latencies if they offer static processing such as equalization (EQ), but dynamics processors and some other types may have much longer latencies, especially if they use look-ahead techniques. RTAS plug-ins mostly do not have significant processing latencies, although some do.

You can easily check the number of samples of delay on any track, whether introduced by a plug-in or by an external effects processor, and then manually correct for this. Just Command-click (Mac) or Control-click (Windows) on the Vol/Peak/Delay view below the track fader in the Pro Tools Mix window until the Delay view appears, and you will see the amount of delay in samples revealed here.

You can manually compensate for this delay by inserting a DigiRack Time Adjuster plug-in on all the other tracks in the session and setting the delay time on all the other tracks to match that of the delayed track. You can choose from any of the following three separate Time Adjuster plug-ins: short, which allows you to delay the audio signal up to 259 samples; medium with up to 2051 samples; and long with up to 8195 samples.

> ### Note
>
> Another way to compensate for latency is to use the "nudge" feature with this set to samples to nudge the delayed track back in time by the same amount as the delay reported by the track's Delay view. This is faster to set up, so it is useful if just one or two tracks need compensation, but the other method is better if lots of tracks need delay compensation because it is easier to see which tracks are delayed and by what amounts by looking at the Delay view then make adjustments to the Time Adjuster plug-ins, rather than nudging lots of tracks around.

Hardware Setup

The Hardware Setup dialog lets you make various settings for your hardware interface or interfaces. On most systems, you will make these settings when you install your Pro Tools system and then leave them alone from that point onwards.

You may need to adjust the hardware setup if you are hooking up additional external equipment for a particular session. For example, on a recent session, I connected a Cedar DNS 2000 Dialog Noise Suppressor to the S/PDIF input and output on my 96 I/O Interface and routed this using Input and Output pairs 9-10 in the Pro Tools software.

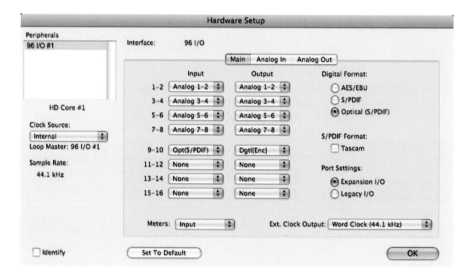

Figure 2.5 Pro Tools|HD Hardware Setup dialog showing Input and Output pairs 9-10 routed via the optical S/PDIF input and output on the 96 I/O Interface.

The Hardware Setup dialog lets you decide which inputs and outputs within the software the physical hardware inputs and outputs will be connected to. In the Hardware Setup dialog, the numbers at the left of the central section (1–2 through 15–16) represent the numbers that will be given to the pairs of inputs and outputs in the Pro Tools software. The columns marked Input and Output have pop-up selectors that let you select which physical inputs or outputs will be routed to the Pro Tools inputs and outputs represented by the numbers to the left. In the example shown, the pop-up selector has been used to select the S/PDIF input that will be routed to Pro Tools input pair 9-10.

I/O Setup

The I/O Setup dialog shows the signal routings for each connected audio interface and allows you to change the routing connections between the physical ports on the audio interface and the inputs and outputs to Pro Tools. You can change the routings from your hardware interface(s) into the Pro Tools software using the row of pop-up selectors just below the interface's icon in the I/O Setup dialog.

> ## Note
>
> These controls mirror the routing controls that you will find in the Hardware Setup dialog, and changes made in the I/O Setup dialog are always reflected in the Hardware Setup dialog.

> ## Tip
>
> If you have opened a Pro Tools session that was recorded on another system, you will probably have to pass through the different tabs for Input, Output, Insert, Bus, and so forth, clicking on the Default button for each I/O Setup section to reset these labels to the defaults for your system.

If you have made any special routings for your Session, it is always a good idea to name the Inputs, Outputs, and any Inserts or Bus routings to make it easier to choose these in the Pro Tools software.

For example, I named Pro Tools input and output pair 9-10 "S/PDIF" to make it clear that these were connected to the S/PDIF I/O on the hardware. With hindsight, it might have been even better to name these as "To Cedar Input" and "From Cedar Output." The point here is that you can use descriptive names for the Inputs, Outputs, Inserts, and so forth, which makes things much easier when you are setting up your tracks and mixer configurations during your Pro Tools session.

Figure 2.6 I/O Setup dialog showing a pop-up selector being used to change the routing from the Pro Tools software to a hardware interface.

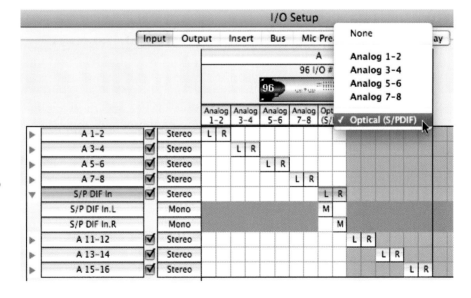

The Output section in the I/O Setup dialog also lets you choose various other settings such as which physical outputs on your interface are to be used as Audition Paths, AFL/PFL path, and AFL/PFL Mute.

Note

AFL is the acronym for After Fade Listen, and PFL is for Pre Fade Listen. PFL is useful for monitoring channels with the faders all the way down so that you can listen for a noisy microphone coming through a channel without hearing this in your mix. You could send the PFL channels to a headphone amplifier, for example, so you can independently monitor these channels to make sure they are OK before feeding them into the mix that you are sending to a PA system or for broadcast. When you monitor using AFL, the volume level of what you hear depends on where the fader is set. So, if you pull the channel fader all the way down you won't hear anything through this channel when you are using AFL.

Auditioning Audio from Other Applications

There will be times when you want to audition audio playing back from applications other than Pro Tools while you are running Pro Tools. The problem with this is that when Pro Tools is running it will not allow any other application to use the audio hardware.

This can be awkward when you want to play iTunes back through your main monitors, for example, or to set up iChat on your Mac for video conferencing over the Internet using your main monitors instead of the Mac's internal speakers. Or you might want to play back some files from another audio application such as BIAS (Berkley Integrated Audio Software) Peak.

One way to handle this is to connect the audio output from your computer's built-in soundcard directly to your monitors. If you are monitoring Pro Tools through an external mixer or some kind of monitor switching system, then you will probably have a stereo input to this that you can use to listen to your computer's audio output at the same time as the Pro Tools audio output.

It can sometimes be more convenient to route the computer's audio output back into Pro Tools using a pair of available inputs on your Avid interface. For instance, I take the optical S/PDIF output from my Mac and connect this to the optical S/PDIF input on my Pro Tools 96 I/O interface. In the Hardware Setup window, I choose optical S/PDIF as the digital format and select a pair of inputs on the interface to use the optical S/PDIF ports.

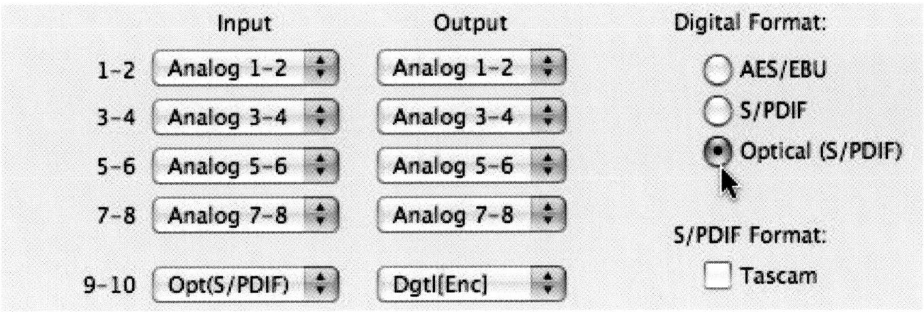

Figure 2.7
Choosing the Optical S/PDIF format in the Hardware Setup window.

In the I/O Setup window, I make sure that the inputs I am using are switched to S/PDIF and named something descriptive such as "Audio from the Mac."

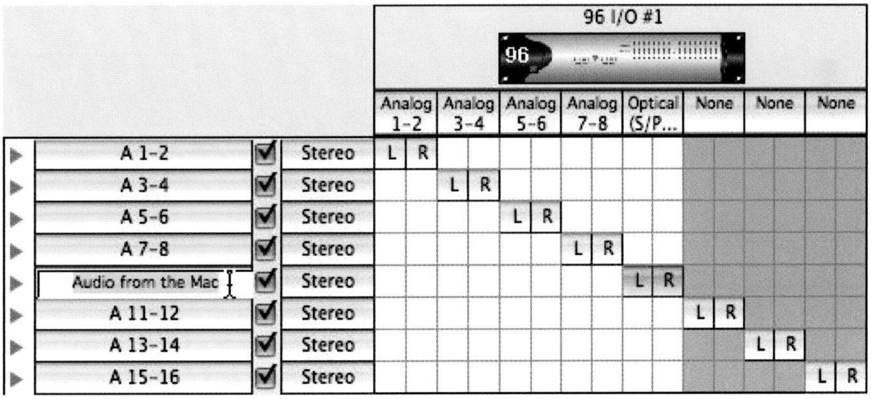

Figure 2.8
Naming an Input in the I/O Setup window.

Don't forget that you will also need to set the sound output from the Mac to use the optical digital outputs.

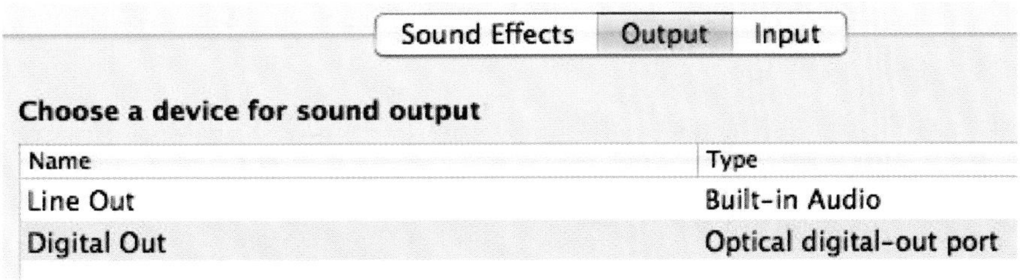

Figure 2.9 Set the Sound Output in the System Preferences to use the optical digital out Port.

You may also need to change the sample rate to match your session using the Mac's Audio MIDI Setup utility.

Note

One limitation here is that the Mac cannot handle sample rates higher than 48 kHz. The practical consequence of this is that you would hear random clicking sounds if you try to play 44.1 or 48 kHz audio from your Mac through an 88.2 or 96 kHz (or higher sample rate) Pro Tools session because the digital audio would not be correctly synchronized.

Figure 2.10
Changing the sample rate in the Audio MIDI Setup window.

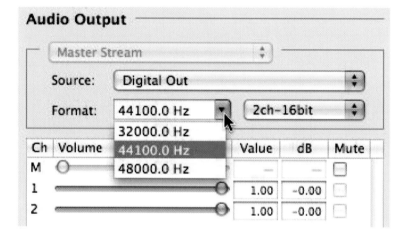

With this set up, all that remains is to insert a stereo Aux channel into your Pro Tools session and choose "Audio from the Mac" as its input to hear audio coming from any of your Mac applications, or from the system, routed into your Pro Tools session.

Tip

With the built-in audio routed into your Pro Tools session through an Aux channel, you will hear any beeps or other system sounds that are produced when you save a file or empty the trash. It can be a problem if you forget about this, especially in the middle of a mix, so it is probably best to mute this Aux channel whenever you are not specifically using it.

Figure 2.11 Aux channel monitoring the Mac's audio.

The User Interface

Pro Tools 9 continues to offer the graphic user interface (GUI) introduced with Pro Tools 8 that offers a more colorful look designed to appeal to creative musicians.

There are very few changes from Pro Tools 8, so anyone upgrading from this version is unlikely to find any real difficulty in making the transition. The main differences are that Pro Tools LE users will find some more advanced features that used to be reserved for HD users are now available to Pro Tools 9 users, especially when using the Complete Production Toolkit option—see Chapter 8 for more details.

The QuickStart Dialog

When you launch Pro Tools, the QuickStart dialog appears (unless you have earlier deselected this option in the Operation Preferences). Using this dialog, you can create a blank session, create a new session from a template, open any of the last ten earlier opened sessions, or open any existing session.

> **Tip**
>
> If you click on the Session Parameters arrow in the QuickStart dialog, you can choose a different audio file type, bit depth, or sample rate for your new session.

A selection of templates is provided in the QuickStart dialog containing useful Session configurations, and you can easily create your own if you prefer. Simply create a new Pro Tools session, configure it however you like, and choose "Save As Template" from the File menu. For songwriting, you might just have one mono track for guitar, a second mono track for voice, a stereo Instrument plug-in for drums or percussion, an Auxiliary Input with a reverb plug-in inserted, and a stereo Master Fader.

In the Save Session Template dialog, you can choose "Install Template In System," which installs the template file in the system folder referenced by the Pro Tools Session Quick Start dialog.

Alternatively, you can choose "Select Location For Template," which lets you select any other location, in which case the session template won't appear in the Pro Tools Session Quick Start dialog—but you can simply open this file to start a new session from this template.

> **Tip**
>
> You might prefer this second method if you want to keep the template with the project that you are working on and intend to move this to another system at some point during the project.

When the Include Media option is enabled, any audio, MIDI, or video media in the session are included in the template. This is useful if you want to use some standard media elements, such as a particular sound effect or drum loop, in a series of related Pro Tools sessions.

Opening Sessions with Plug-ins Deactivated

If you are using a lot of plug-ins in a particular session, these can take a long time to load. If you simply want to open the session to check if this is the one you want to work with, or to make some changes that you will not need the plug-ins for, Pro Tools provides a convenient way to open sessions with all of the session's plug-ins set to inactive.

From the File Menu, choose Open. When the Open Session dialog appears, first locate and select the session you want, and then Shift-click Open. The Session will now open with the plug-ins deactivated.

If you decide that you do want to work with this file with the plug-ins loaded and activated, you can either choose Revert To Saved from the File menu or choose Open Recent from the File menu and select the most recent session in the submenu.

> ### Tip
> There is a convenient keyboard command that lets you reopen the most recently opened session: Command-Shift-O (Mac) or Control-Shift-O (Windows).

Setting the Color Palette

One of the first things you will notice if you have used earlier versions of Pro Tools is the greater use of color in the new user interface. The Color Palette lets you make color selections for tracks, regions, groups, and markers. It also lets you apply colors to channel strips in the Mix and Edit windows, and lets you adjust the saturation and brightness of the colors to emulate the way that other popular user interfaces look.

Managing Your Window Configurations

Pro Tools lets you save your current arrangement of windows on-screen as a "window configuration" that you can quickly recall at any time from the Window Configurations List.

To open the Window Configurations List, you can either select this from the Configurations submenu in the Window menu or press Command-Option (Mac) or Control-Alt (Windows) and the letter "j" on your computer keyboard. Here, you will see any window configurations that you have already created.

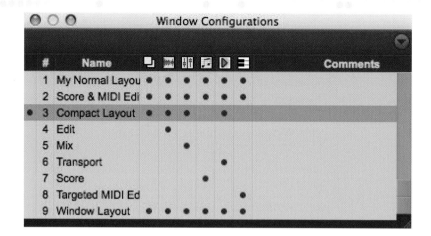

Figure 2.12 The Window Configurations List showing nine saved configurations.

Only one window configuration containing a Window Layout can be active at any one time, and to indicate which of the configurations in the list is active, this configuration will have a black dot to the left of its list number.

If you have created lots of window configurations, it can be difficult to find the one you want to recall quickly. To make this easier, you can either filter out configurations that you are not interested in using the options in the Window Configurations List pop-up menu or click on the View Filter icons in the Window Configurations List.

Using either of these methods, you can show or hide window configurations based on whether or not they are stored with Window Layout, Edit Window settings, Mix Window settings, Score Editor, Transport, or Targeted MIDI Editor Window settings.

Figure 2.13 Window Configurations List window showing pop-up menu with view filter options.

If you want to change the way you have configured one of your window configurations, just select it, make your changes, then choose the Update "configuration name" command from the Window Configurations List pop-up menu.

> ### Tip
> I find myself mostly working within one "normal" layout that shows both the Edit and Mix windows. However, as I add tracks to the mixer and adjust the size of the Mix window and change settings in the Edit window, I usually want to retain these new layouts and settings. The best way to do this is to enable Auto-Update Active Configuration in the Window Configurations submenu.

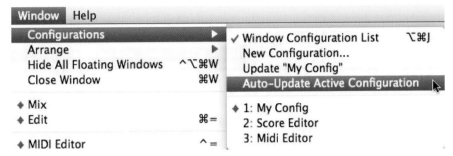

Figure 2.14
Auto-Update Active Configuration option in the Configurations submenu.

Creating Window Configurations

Why would you want to use different window configurations? Well, for example, maybe you want to close the Mix and Edit windows when you are editing Score and MIDI data so that you can focus on using just these windows. All you need to do is to close the windows that you don't want and open the ones you do want, positioning these wherever you like on-screen.

Figure 2.15
Score and MIDI Editing window configuration.

To create the configuration, choose "New Configuration" from the pop-up menu available at the right of the Window Configurations List. In the New Window Configuration dialog that appears, you can name your configuration then click "OK" to save it.

If you enable the Window Layout button, the size and location of all the open windows will be stored with the window configuration. A checkbox below this lets you optionally include all the window display settings (such as whether or not the Region List is shown in the Edit window) for the Edit, Mix, Targeted MIDI Editor, Score Editor, and Transport Display.

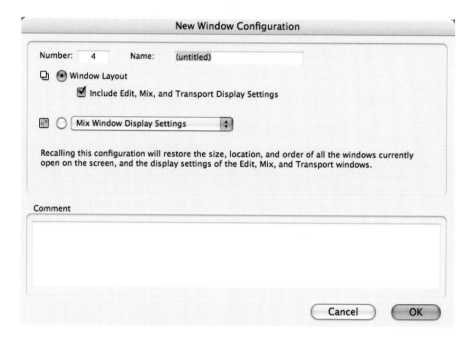

Figure 2.16
Creating a
new window
configuration.

You can store and recall up to 99 window configurations along with other settings such as the Zoom Toggle preferences, the track heights, and the internal window configurations of the main windows.

> **Tip**
>
> The window configuration memory slot numbers can be used to quickly recall individual configurations: using the numeric keypad on your computer's keyboard, press Period (.) first; release this, and then press the number of the window configuration; release this, and then press the Asterisk (*). This series of actions will recall the window configuration stored in that slot.

Window configurations are particularly useful when you have limited screen space, so I always create a compact layout containing the Edit and Mix windows, for example, when I am using a laptop.

Window Display Settings

You can also store the window display settings for the Edit, Mix, Transport, Score Editor, and Targeted MIDI Editor windows as recallable window configurations.

The window display settings for the Edit window include the width of the Tracks List and Group List; the height of the Tracks List; the width of the Region List; what rulers are shown (the main ruler is always shown); what track columns are shown (such as Inserts, Sends, or Comments); the Tempo editor display; and whether the Transport controls are shown in the Edit window.

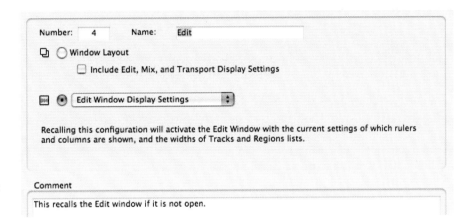

Figure 2.17 Edit Window Display Settings.

The window display settings for the Mix window include the width of the Tracks List and Group List, the height of the Tracks List, what track rows are shown (such as Inserts, Sends, or Comments), and the Narrow/Wide mixer view.

The Transport Window Display Settings option stores all window display settings for the Transport window, including the Counters, MIDI controls, and Expanded view settings, and lets you recall the Transport window with these settings when it is closed.

The Score Editor Window Display Settings option stores all the current window display settings for the Score Editor window.

The Targeted MIDI Editor Window Display Settings option stores the current automation panes and heights, and the current settings of which rulers and rows are being shown in the targeted MIDI Editor window.

> **Tip**
>
> If you just want to to recall the size, location, and order of a particular set of windows that you like to work with, but without recalling the display settings within those windows, you can simply store the Window Layout with the "Include … Display Settings" option deselected in the New Window Configuration dialog.

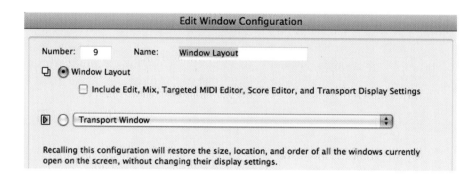

Figure 2.18 Window Layout configuration.

The Universe View

Pro Tools has a navigation feature called the Universe View that runs along the top of the Edit window. This displays an overview of the entire session, showing a compact representation of all the audio and MIDI data on tracks that are not hidden.

Audio, MIDI, and video regions on tracks are represented by horizontal lines that are of the same colors as the regions on the tracks, and each channel in a stereo or multichannel audio track is represented individually. Since Auxiliary Input, Master Fader, and VCA Master tracks do not contain audio or MIDI regions, they are displayed as blank areas in the Universe View.

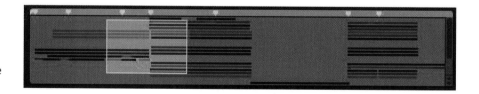

Figure 2.19 The Universe View.

To make a material visible within the confines of the Edit window, you often need to use the vertical and horizontal scroll bars so that you can see what you need to see. The Universe View provides an alternative way of positioning material in the Edit window so that it is visible.

A white rectangular area within the Universe View represents what is visible in the Tracks area of the Edit window. You can click on and drag this white rectangle to move the area that it encompasses (or "frames"), and the material displayed in the Edit window will follow. This can be a much faster method than using the scroll bars to achieve the same result.

The framed area updates to reflect what is happening in the Edit window, so when all the tracks are made visible in the Edit window, and the session is zoomed all the way out with all regions visible in the Edit window, the framed area encloses the whole of the Universe window. During playback, if the Edit window is set to scroll, the framed area in the Universe View will scroll to keep in step with this.

Pro Tools also has a Video Universe window accessible from the Window menu that lets you view, navigate, zoom, and select video regions on the main video track.

The Main Windows

The Edit and Mix windows are the main Pro Tools work areas. Depending on which phase of your project you are in or what type of project you are working on, you may prefer to work with just the Mix or just the Edit window.

You can use the View Menu to choose what will be presented to you in the Edit and Mix windows. So, for example, if you reveal the Instruments, I/O controls, Inserts, and Sends in the Edit window, you can mostly work in just the Edit window, which some people find easier than working in two windows.

Figure 2.20
Revealing the I/O controls in the Edit window.

Organizing the Transport Window

The Transport window takes its name from the name given to the controls that were used on tape recorders to control the tape transport, that is, the mechanism that caused the tape to be transported along its path across the tape heads. Presumably, Avid chose this name to make recording engineers and producers who started out using tape recorders feel more comfortable with their digital audio recorder. Interestingly, a hard disk drive is like a cross between a tape recorder and a vinyl disc recorder: like a vinyl disc recorder it has a rotating platter and a read/write head on a movable arm (or several of these), and like a tape recorder, it records the audio as patterns of magnetization on a magnetic medium—an analog process. But as far as the computer is concerned, the audio is recorded and played back digitally as a series of 1's and 0's from a file on disk.

Clearly, the Transport window is an important part of any Pro Tools—allowing you to control recording and playback of audio (and MIDI)—so you need to make sure that you are totally familiar with all its controls and functions.

In standard view, the Transport window just has a row of buttons providing controls for Online, Return To Zero, Rewind, Stop, Play, Fast Forward, Go to End, and Record. The Transport submenu that you can access from within the View menu lets you display Synchronization controls, MIDI Controls, and Counters in the Transport window as well. Synchronization controls include two buttons: one to enable Online timecode synchronization and the other to generate MIDI Time Code (MTC) using a SYNC peripheral.

There are four MIDI control buttons: Wait for Note, Metronome, MIDI Merge, and Conductor. The Wait for Note does what it says: Pro Tools won't play or record until you send it a MIDI note. This is useful when you are controlling Pro Tools yourself and are playing a MIDI instrument such as a keyboard. The Metronome button enables the click, assuming you have set up a click track correctly. The MIDI Merge lets you record new MIDI data onto a track that already contains MIDI data, merging all the data together. The Conductor button lets you enable or disable the Conductor track that contains tempo information for your Pro Tools session.

The Main Counter shows your choice of Bars:Beats, Mins:Secs, Samples, Timecode, Timecode 2, or Feet + Frames, and you can choose between these using the pop-up selector that appears when you click and hold the small downwards pointing arrow at the right of the counter display.

You can also switch the Transport window to its Expanded display. This adds a Subcounter below the main counter, along with Count Off, Meter, and Tempo controls above the MIDI controls. You can either enter the tempo manually here or enable the Tempo track by clicking on the Conductor button.

Below the transport controls, the Expanded Transport window adds Pre and Postroll settings, along with Start, End, and Length indicators for making Timeline selections.

Figure 2.21 Transport Window expanded to show the Counters and MIDI Controls with the pop-up selector for these revealed at the far right.

Tip

To start and stop playback, simply press the Spacebar on your computer keyboard.

Transport Preferences

To be fully in control of Pro Tools, you need to understand the effects that the various Transport Preferences settings have.

If "Timeline Insertion/Play Start Marker Follows Playback" is selected, when you stop playback the Timeline Insertion and the Play Start Marker move to the point on the Timeline where playback stops. Otherwise, when this option is deselected, the Timeline Insertion and Play Start Marker will return to the point on the Timeline from which playback began.

Note

This is such an important setting that there is a dedicated button on the Toolbar, marked "Insertion Follows Playback," that lets you toggle this option on and off.

When "Edit Insertion Follows Scrub/Shuttle" is selected, the edit cursor automatically locates to the point on the Timeline where you stop scrubbing the audio.

When "Audio During Fast Forward/Rewind" is selected, you will hear the audio playing back during fast forward or rewind.

When "Latch Forward/Rewind" is selected, Fast Forward and Rewind will latch and continue to operate until you start or stop playback. When this preference is disabled, the Fast Forward and Rewind will only last as long as you hold down the mouse after clicking either button on the Transport or hold down the corresponding switch on a Control Surface.

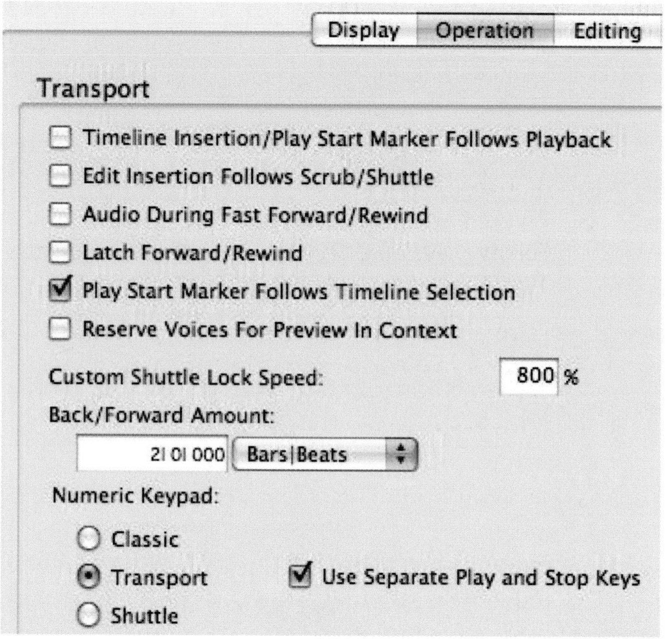

| Display | Operation | Editing |

Transport

☐ Timeline Insertion/Play Start Marker Follows Playback
☐ Edit Insertion Follows Scrub/Shuttle
☐ Audio During Fast Forward/Rewind
☐ Latch Forward/Rewind
☑ Play Start Marker Follows Timeline Selection
☐ Reserve Voices For Preview In Context

Custom Shuttle Lock Speed: [800] %
Back/Forward Amount:
[2I 0I 000] [Bars|Beats ▲▼]

Numeric Keypad:
○ Classic
⦿ Transport   ☑ Use Separate Play and Stop Keys
○ Shuttle

Figure 2.22 Pro Tools HD Transport Preferences.

When "Play Start Marker Follows Timeline Selection" is enabled, the Play Start Marker snaps to the Timeline Selection Start Marker when you move the Timeline Selection or when you draw a new Timeline Selection or when you adjust the Timeline Selection Start. When this preference is disabled, the Play Start Marker doesn't move with the Timeline selection.

Pro Tools HD systems have an extra option in the Transport Preferences: "Reserve Voices for Preview in Context."

When "Reserve Voices For Preview In Context" is enabled, Pro Tools|HD reserves the appropriate number of voices for preview in context (the feature that allows previewing audio files in DigiBase during session playback).

> **Note**
>
> When you are reserving voices for Preview in Context, these are no longer available to use for other purposes because the number of available voices is inevitably reduced by the channel width of the selected Audition Paths on the Output page of the I/O Setup window. For example, if the number of playback voices is set to 48 in the Playback Engine, and you have a stereo audition path selected in the I/O Setup, only 46 voices will be available for tracks. If you have a 5.1 audition path, only 42 voices will be available. When this option is disabled, you will not be able to preview in context if there are not enough available voices.

The Custom Shuttle Lock Speed preference (Pro Tools HD and Complete Production Toolkit 2 only) allows you to customize the highest fast-forward Shuttle Lock speed in Transport or Classic numeric keypad modes. The range for this setting is 50–800%.

You can use the Back/Forward Amount to set the default length of Back, Back and Play, Forward, and Forward and Play. The timebase of the Back/Forward Amount settings follows the Main Time Scale by default, or you can deselect Follow Main Time Scale and select another timebase format: Bars:Beats, Min:Sec, Time Code, Feet+Frames, or Samples.

When "Use Separate Play and Stop Keys" is enabled, you can start playback with the Enter key and stop playback with the Zero (0) key on the numeric keypad. The "Use Separate Play and Stop Keys" is only available when you are using the Numeric Keypad's Transport mode.

Tip

Setting up Pro Tools to use the Enter and '0' keys on the numeric keypad to start and stop playback makes Pro Tools behave like Cubase and other popular MIDI + Audio software applications and is particularly useful when you want to quickly audition loop transitions.

Note

When "Use Separate Play and Stop Keys" is enabled, this option overrides using the Enter key to add Memory Location markers, so if you do need to add a Memory Location marker while this is active, you need to press Period (.) before you press Enter on the numeric keypad.

Scrub/Shuttle and the Numeric Keypad Modes

To understand which Numeric Keypad mode to choose, you need to understand the Scrub/Shuttle features first.

Using the Scrubber Tool

Scrubbing is a technique borrowed from tape editing, where the tape was moved back and forth past the playback head by hand while listening to the audio to find a suitable edit point for cutting or splicing. The Scrubber tool in Pro Tools emulates this procedure, allowing you to "scrub" up to two tracks of audio in the Edit Window (MIDI and Instrument tracks cannot be scrubbed).

To scrub a single audio track, select the Scrubber tool then point, click and drag within the track, moving left for reverse or right for forward.

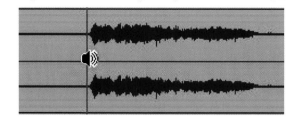

Figure 2.23
Scrubbing by
dragging within a
single track.

To scrub two audio tracks, drag between the adjacent tracks or scrub within a selection that contains multiple tracks.

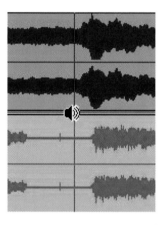

Figure 2.24
Scrubbing by
dragging between
two adjacent
tracks.

If the "Edit Insertion Follows Scrub/Shuttle" option is enabled in the Operation Preferences page, when you stop scrubbing, the edit cursor will automatically locate to the point where you stopped. It makes most sense to enable this option when you are using the Scrubber tool because you will usually want to insert the edit cursor at this point to make your edit.

To make the scrub feature even more like tape editing, you can set the Scrolling option to Continuous or Center Playhead. With either of these scrolling modes, wherever you click in a track's playlist with the Scrubber tool activated, Pro Tools scrolls so that the Playhead is centered in the Edit window. Then, as you scrub the audio, it moves past the Playhead, which remains stationary and centered—more like moving tape back and forth past a fixed play head.

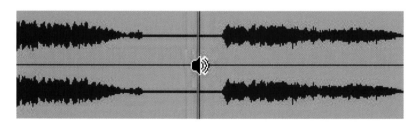

Figure 2.25
Scrubbing with
Center Playhead
scrolling enabled.

To get used to using the Scrubber tool is easy enough. Just move the mouse back and forth over the audio where you are trying to find the correct edit point, moving progressively more slowly until you have identified the spot. This can be much easier to control if you are using a scrub wheel on a control surface.

Keep in mind that the speed of playback of the scrubbed audio depends on how fast and how far you drag. Also, the resolution for the Scrubber is dependent on the zoom factor for the scrubbed track, so you may need to zoom in or zoom out for best results.

> **Tip**
>
> You can temporarily switch the Selector tool to the Scrubber tool by Control-clicking (Mac) or Start-clicking (Windows). For finer resolution, Command-Control-click (Mac) or Control-Start-click (Windows).

Shuttle Mode

Shuttle mode is used for Pro Tools HD and Complete Production Toolkit 2 only.

The Scrubber tool lets you scrub at normal playback speeds or slower.

Sometimes when you are looking through larger ranges of audio, you may want to scrub at faster than normal speeds ,which is called Shuttle mode.

To scrub in Shuttle mode, simply hold down the Option key (Mac) or Alt key (Windows) while you drag within the track with the Scrubber tool selected. The Fast Forward and Rewind buttons in the Transport window will engage to indicate that you are in fast forward/rewind Shuttle mode.

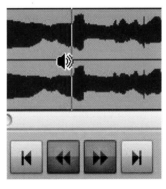

Figure 2.26
Scrub/Shuttle mode showing the Fast Forward and Rewind buttons in the Transport window engaged.

Shuttle Lock Mode

Shuttle Lock mode lets you shuttle forwards or backwards through your audio while controlling the speed and direction using the computer keyboard instead of using the mouse or scrub wheel.

To enter Shuttle Lock mode, hold down the Control key (Mac) or Start key (Windows) and press a number on the numeric keypad. You can let go of the modifier key once playback has started, and you can reverse the direction of playback by pressing the Minus (–) or Plus (+) keys on the numeric keypad.

To change the speed of playback, hold down the modifier key again and press a different number on the numeric keypad: 5 is normal speed, numbers from 6 up to 9 provide increasingly faster fast-forward speeds, and numbers from 4 down to 1 provide progressively faster rewind speeds (4 is the slowest rewind Shuttle Lock speed, 1 is the fastest).

To stop playback, press Control-0 (Mac) or Start-0 (Windows), and to exit Shuttle Lock mode, just press Escape or the Spacebar. Remember that, as with all the Scrub/Shuttle modes, if multiple tracks are selected, only the first two tracks will be heard.

The Numeric Keypad Shuttle Lock Modes

There are two numeric keypad Shuttle Lock modes: Classic and Transport.

Classic mode emulates the way Pro Tools worked in versions lower than 5.0. To control Shuttle speed, press the Control key (Mac) or Start key (Windows), followed by a number from 0–9 for the different play speeds. In Classic mode, you can recall Memory Locations (markers) by simply typing the Memory Location number, followed by Period (.)—using your computer's numeric keypad.

As with Classic mode, Transport mode allows you to control Shuttle speed by pressing the Control key (Mac) or Start key (Windows), followed by a number from 0–9 for the different play speeds. The difference is that Transport mode also allows you to set a number of record and play functions and lets you operate the Transport using the numeric keypad.

With Transport mode active, you can press 0 to Play/Stop; 1 to Rewind; 2 to Fast Forward; 3 to enable Record; 4 to turn Loop Playback mode on/off; 5 to turn Loop Record mode on/off; 6 to turn QuickPunch mode on/off; 7 to turn the Click on/off; 8 to turn the Countoff on/off; and 9 to enable MIDI Merge/Replace mode.

You can recall Memory Locations (markers) in Transport mode by typing Period (.), Memory Location number, and Period (.) again—using your computer's numeric keypad.

The Numeric Keypad Shuttle Mode

The numeric keypad Shuttle mode provides an alternative to use Shuttle Lock mode when you want to control Scrub/Shuttle using the computer keyboard instead of the mouse/scrub wheel.

Shuttle mode works quite differently from the two Shuttle Lock modes. The main difference is that to play the current Edit selection you must press and hold the keys on the numeric keypad; playback stops as soon as you let go of the keys. And you don't have to be in Scrub mode to use the Shuttle feature because Scrub mode is automatically engaged as soon as you press the relevant keys on the numeric keypad. Also, in this mode, pre and postroll are ignored.

Various playback speeds are available in both forward and reverse. Press key 1 to Rewind at quarter speed; key 3 to go Forward at quarter speed; key 4 to Rewind at normal speed; key 6 to go Forward at normal speed; key 7 to Rewind at 4x speed; and key 9 to go Forward at 4x speed. You can Loop a Selection at normal speed by pressing key 0.

For more options, you can use pairs of keys: press keys 1+2 to Rewind at 1/16x speed; keys 2+3 to go Forward at 1/16x speed; keys 4+5 to Rewind at half speed; keys 5+6 to go Forward at half speed; keys 7+8 to Rewind at double speed; and keys 8+9 to go Forward at double speed.

You can recall Memory Locations in Shuttle mode by typing Period (.), Memory Location number, and Period (.) again—using your computer's numeric keypad.

Note

No matter which Numeric Keypad mode is selected, you can always use the numeric keypad to select and enter values in the Event Edit Area, Edit Selection indicators, Main and Sub Counters, and Transport fields.

Organizing the Mix Window

In the Mix window, tracks appear as mixer channel strips with controls for signal routing, input and output assignments, volume, panning, record enable, automation mode, and solo/mute.

From the View menu, you can choose whether or not to display the Mic Preamps, Instruments, Inserts, Sends, Delay Compensation, and Track Color controls, or Comments in the Mix window.

The Group Settings Pop-up Selector

Although most of the Pro Tools 8 user interface was very similar to earlier versions in terms of layout and features, there were various small changes intended to make things clearer or easier to use. Pro Tools 9 maintains these features.

For example, directly above the Pan controls at the top of each fader section, there is a pop-up selector that lets you control any Group settings that you have made for the track.

Figure 2.27 Pro Tools HD Mix window with mouse pointing to the Group settings pop-up selector. The pop-up selector for an adjacent track is open in front of the Mix window.

Using this pop-up selector, you can select or hide all the tracks in the group, show only the tracks in the group, delete, duplicate, or modify the group, and see which tracks and attributes are included in the group.

Inserts

Pro Tools lets you use up to ten Inserts on each Audio track, Auxiliary Input, Instrument Track, or Master Fader. Each insert can be either a software plug-in or an external hardware device. There are two sets of Inserts, labeled A–E and F–J, so you can conserve space on-screen by only displaying one set at a time.

Sends

You can also use up to ten Sends on each track. Sends let you route signals across internal buses or to audio interface outputs so that one plug-in or one external signal processor can be used to process several tracks at once.

There are two sets of Sends (labeled A–E and F–J) that can optionally be displayed in the Mix window. You might use the first set to send effects (such as reverb) that you wish to apply to several tracks, and use the second set to send cue mixes to musicians—routing these from your Pro Tools hardware interface to suitable headphone amplifiers.

If you are not using the second set of Sends, or the mic preamps, or you don't need to see the delay compensation controls or comments, or whichever, then you should hide these so that they don't distract you from the controls you do want to use and so that the Mix window takes up less space on your computer screen.

I/O Selectors

Track Input and Output Selector pop-ups are located below the Inserts and just above the Automation controls on the channel strips. The Track Input Selector pop-ups let you choose input sources for Audio tracks, Instrument tracks, and Auxiliary Inputs. Track input can come from your hardware interface, from an internal Pro Tools bus, or from a plug-in.

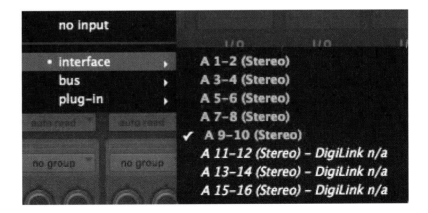

Figure 2.28 Track Input Selector pop-up.

The Track Output Selector pop-ups let you route the audio from each track to your choice of available outputs or bus paths.

Assigning Multiple Track Outputs and Sends

Pro Tools Audio tracks, Instrument tracks, and Auxiliary Inputs can have multiple track output and send assignments chosen from the actual paths and resources available on your system (although Master Faders can only be assigned to a single path).

Assigning to multiple paths is an efficient way to route an identical mix to other separate outputs, for simultaneous monitor feeds, headphone mixes, or other situations where a parallel mix is needed.

To assign an extra output, hold down the Control key (Mac) or Start key (Windows), open the Output Selector, and select your additional output.

A "+" sign is added to the Output Selector legend to remind you that this track has more than one output assigned, and you can add as many additional outputs available on your system.

If you also hold the Option (Alt) key at the same time as the Control (Start) key, the additional output will be added to all tracks (apart from Master Faders and MIDI tracks, of course).

Figure 2.29
Making multiple
assignments for
a track's Audio
Output Path.

> ### Note
>
> You can use the same procedure to add additional output assignments to the track Sends.

Organizing the Edit Window

The Edit window provides a timeline display of audio, MIDI data, and mixer automation. As in the Mix window, each track has controls for record enable, solo, mute, and automation mode.

Using the View menu options, you can also reveal the input and output (I/O) routing assignments, the Inserts, the Sends, the Instrument controls, the Real-Time properties, the Comments, or any combination of these, which makes it possible to work with just the Edit window for most of the time.

Figure 2.30
Multiple
assignments for
a Send's Audio
Output Path.

The Edit window lets you display the audio and MIDI data in a variety of ways to suit your purpose, and you can edit the audio right down to sample level in this one window.

The Toolbar

Understanding the area at the top of the Edit window is an essential part of learning how to use Pro Tools. This area, called the Toolbar, contains the Edit Mode buttons, the Zoom buttons, the Edit tools, the Location indicators, various other displays and controls, and the Grid and Nudge controls.

In Pro Tools 9, you can customize the Toolbar in the Edit window (as with the MIDI and Score Editor windows) by rearranging, showing, and hiding the available controls and displays. To show or hide the various tools, you can use the pop-up Toolbar menu located in the upper right corner of the Edit window.

Here, you can choose to display the Transport, Expanded Transport, MIDI, and Synchronization controls in the Toolbar if you wish. You can also show or hide the Track List or Regions List, reveal the Universe View at the top of the Edit window, or reveal the MIDI Editor at the bottom of the Edit window.

With all these showing in the Edit window, this allows you to work almost entirely within the one window. The disadvantage is that the Edit window then starts to look very cluttered—that is why you are given the option to hide whichever of these elements you are not using.

Figure 2.31 Toolbar showing the basic set of controls and displays.

Rearranging Toolbar Controls and Displays

To rearrange controls and displays in the Edit (or MIDI or Score Editor) window toolbar, simply Command-click (Mac) or Control-click (Windows) on the controls or displays that you want to move and drag them to the right or to the left to reposition them along the toolbar. So, for example, if you prefer to have the Counters and Edit Selection indicators repositioned to the right of the Transport controls in the toolbar, it will only take you a moment to do this.

The Edit Modes

At the left of the Toolbar, there are four buttons to let you select the Edit mode:

Figure 2.32 The Edit Mode buttons.

- Slip mode is the basic mode to use by default. In this mode, you can freely move regions forward and backward in time in the Edit window.
- In Shuffle mode, when you move a region, it will automatically snap to the region before it—perfectly butting-up to this.

- In Spot mode, a Spot dialog appears whenever you click on a region. This lets you specify exactly where the region should be placed on the timeline—ideal for "spotting" effects to picture.
- In Grid mode, movements are constrained by whichever Grid settings you have made—such as Bars:Beats or Mins:Secs.

Pro Tools lets you Shift-click to enable Grid mode while in Shuffle, Slip, or Spot mode. You can also use the keyboard command Shift-F4 if you prefer.

With Grid mode active, placing the Edit cursor and making Edit selections are constrained by the Grid, while region editing is simultaneously affected by the other selected Edit mode. For example, in Shuffle mode, with Grid mode also active, you can make a selection in a region based on the Grid, cut the selection, and any regions to the right of the edit will shuffle to the left.

Tip

To enable Shuffle, Slip, or Spot while in Grid mode, Shift-click the Shuffle, Slip, or Spot mode button. Alternatively, press F1+F4 to enable Grid and Shuffle mode; press F2+F4 to enable Grid and Slip mode; and press F3+F4 to enable Grid and Spot mode.

The Toolbar Zoom Buttons and Presets

To the right of the Edit Mode buttons, there are various arrow buttons that let you zoom the display vertically or horizontally. The horizontal zoom arrows work for both audio and MIDI regions. To zoom audio regions vertically, use the first pair of up and down arrow buttons. To zoom MIDI regions vertically, use the second pair.

Figure 2.33
Zoom buttons and presets.

Below these zoom controls there are five small Zoom Preset buttons that store preset zoom levels. You can use these as handy shortcuts to particular zoom levels: just set the zoom level you want, and then Command-click (Mac) or Control-click (Windows) on any of the five buttons to store the current zoom level.

To recall these zoom presets, you can either click on the numbered buttons using the mouse or press the numbers 1, 2, 3, 4, or 5 on your computer's QWERTY keyboard (not on the numeric keypad).

The Edit Window Zoom Buttons

In addition to the Zoom controls in the Toolbar, Pro Tools provides horizontal and vertical zoom buttons in the lower right corner of the Edit window. The Vertical Zoom buttons zoom the track heights proportionally in the Edit window. The Horizontal Zoom buttons zoom the Timeline, just like the Horizontal Zoom controls in the Edit window toolbar.

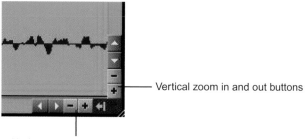

Vertical zoom in and out buttons

Figure 2.34
Horizontal and vertical zoom buttons.

Horizontal zoom in and out buttons

Pro Tools 8 also provides Audio and MIDI Zoom In and Out buttons in the upper right corner of the Edit window. These controls function exactly the same as the Audio and MIDI Zoom controls in the Toolbar and let you zoom in and out vertically on audio waveforms and MIDI notes, respectively, although MIDI Vertical Zoom only affects tracks in Notes view.

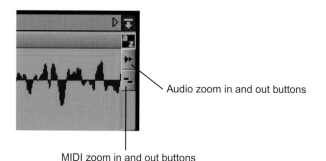

Audio zoom in and out buttons

Figure 2.35
Audio and MIDI Zoom In and Out buttons.

MIDI zoom in and out buttons

The Tool Buttons

Located by default to the right of the Zoom buttons, you will find six Tool buttons.

Zoomer Tool

The first button is the Zoomer Tool. You can use this to zoom the display either vertically or horizontally. If you click and hold this, you can select the Single Zoom mode. With this selected, you can click in the Edit window to zoom the display one time, and then when you let go of the mouse, the Zoomer Tool is deselected, and the Selector Tool becomes active.

Trimmer Tool

The next button is the Trimmer tool. You can use this to lengthen or shorten regions. You can also choose this tool to apply Time Compression/Expansion directly in the Edit window or to apply the Scrub feature prior to trimming regions or to apply the Loop feature. Select these modes using the pop-up that appears when you click on this tool.

Selector Tool

To the right of the Trimmer tool is the Selector tool that lets you use the cursor to select areas within the Edit window. With this tool selected, the mouse pointer changes to an insertion cursor that you can use to drag across and select regions in the Edit window, or you can simply point and click at a particular location in the Edit window to position the insertion point (and update the Location indicators) at that location.

> **Tip**
>
> Whenever you are not specifically using one of the other tools, the Selector tool is the most useful (and safest) tool to leave selected.

Grabber Tool

To the right of the Selector tool you will find the Grabber tool—the one with the "hand" icon. You can use this to move regions around in the Edit window. You can also use the Grabber to automatically separate an edit selection and move it to another location or another track using its Separation mode.

The Grabber tool offers a third mode—the Object mode—that you can enter using the pop-up that appears when you click on the Grabber tool. This "Object Grabber" lets you select noncontiguous regions on one or more tracks. Just take a look at the screenshot to see an example of a "noncontiguous" selection of three regions, all on different tracks. By the way, "noncontiguous regions" in this context basically means regions that are not next to each other.

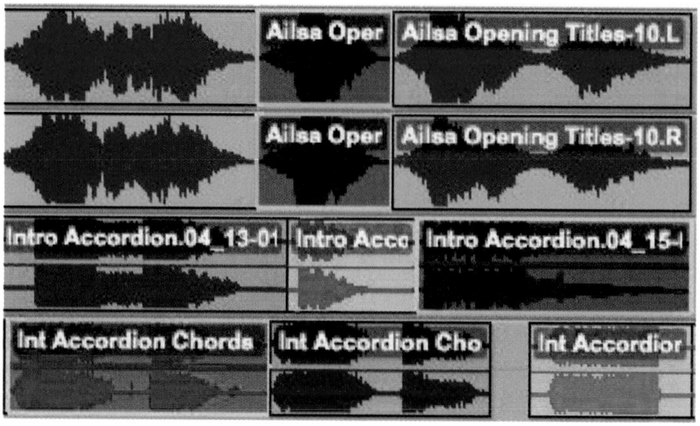

Figure 2.36 The Object Grabber Tool being used to select noncontiguous regions in the Edit window.

Smart Tool

The Trimmer, Selector, and Grabber tools can be combined using the Smart tool to link them together. The Smart Tool button has a new look and location in Pro Tools 9. It is now positioned both above and to either side of the Trimmer, Selector, and Grabber tools. To activate it, simply click in this area above or to either side of these tools. With the Smart tool activated, and then depending on where you point your mouse in the Edit window, one or other of these tools will become active, saving you from having to click on these tools individually when you want to change to a different tool.

Figure 2.37 The Smart Tool.

Scrub Tool

To the right of the Grabber tool is the Scrub tool that lets you "scrub" back and forth over an edit point while you are trying to hear the exact position of a particular sound—rather like moving a tape back and forth across the playback head in conventional tape editing.

Pencil Tool

The sixth tool is the Pencil that you can use to redraw a waveform to repair a pop or click. Alternate Pencil modes are available that constrain drawing to lines, triangle, square, or random shapes.

The Mode Buttons

Present below the Tool buttons, there are six more mode buttons.

Zoom Toggle

The leftmost button controls the Zoom Toggle. This feature lets you define and toggle between zoom states in the Edit window, storing and recalling the Vertical Zoom, Horizontal Zoom, Track Height, Track View and Grid settings. When Zoom Toggle is activated, the Edit window changes to the stored zoom state, and when Zoom Toggle is disabled, the Edit window reverts to the earlier zoom state. Any changes made to the view, while Zoom Toggle is enabled, are also stored in the zoom state.

Figure 2.38
Zoom Toggle
button.

> **Tip**
> When Zoom Toggle is enabled, you can cancel it and remain at the same zoom level by pressing Option-Shift-E (Mac) or Alt-Shift-E (Windows).

Tab to Transients

The Tab to Transients button lets you automatically locate the cursor to the next transient while editing waveforms. With this activated, just press the Tab key on your computer keyboard to jump to the beginning of the next transient (e.g., at the start of a snare hit) in the waveform that you are currently editing.

Figure 2.39 Tab
to Transients
button.

Mirrored MIDI Editing

The third button from the left lets you enable or disable Mirrored MIDI Editing. Mirrored MIDI Editing is useful when you edit a region containing MIDI notes and you want these edits to apply to every MIDI region of the same name.

Figure 2.40
Mirrored MIDI
Editing button.

Link Timeline and Edit Selection

The Link Timeline and Edit Selection button let you link or unlink Edit and Timeline selections. With the Timeline and Edit selections *unlinked*, you can select different locations in the Timeline at the top of the Edit window and in the regions within the Edit window. With the Timeline and Edit selections *linked*, whenever you make a selection inside the Edit window, the same selection is automatically made in the Timeline at the top of the Edit window.

Figure 2.41 Link Timeline and Edit Selection button.

Link Track and Edit Selection

The Link Track and Edit Selection button also does what it says: with this highlighted, when you select a region in the Edit window, the track becomes selected as well. If you then select another track, Pro Tools selects the corresponding region (to the region selected in the first track) in this other track—since the Track and Edit selection features are linked!

Figure 2.42 Link Track and Edit Selection button.

Insertion Follows Playback

Pro Tools 8 provides a new Insertion Follows Playback button in the Edit window. This lets you enable or disable the "Timeline Insertion/Play Start Marker Follows Playback" option and also provides a visual indication of whether or not this option is on. Earlier, this option was only available in the Operation Preferences.

Figure 2.43 Insertion Follows Playback button.

> **Tip**
> There is a useful keyboard command that lets you switch this preference on and off: press Control-n (Mac) or Start-n (Windows) or, with the Commands Keyboard Focus enabled, simply press the "n" key.

Location Indicators

Located centrally at the top of the Edit window you will find the Main and (optional) Sub Location indicators with the Event Edit area to the right of these.

The Location indicators show you where you are in your session in terms of Bars:Beats, Mins:Secs, or whatever you have chosen to display here.

The Event Edit Area lets you define selections, for example, by typing the Start and End points, and also serves to display these along with the length of the selection.

Figure 2.44
Location Indicators
and Event Edit
Area.

Edit Window MIDI Editing Controls

To allow convenient editing of selected single notes when you are working with MIDI or Instrument tracks, Pro Tools 8 provides various MIDI editing controls in the Edit window toolbar.

If you select a MIDI note using the Grabber or Pencil tools, the MIDI Editing Controls area at the right of the Event Edit area displays the selected MIDI note with its associated MIDI On and Off velocities. You can edit these values by typing new values or by dragging the cursor to change values.

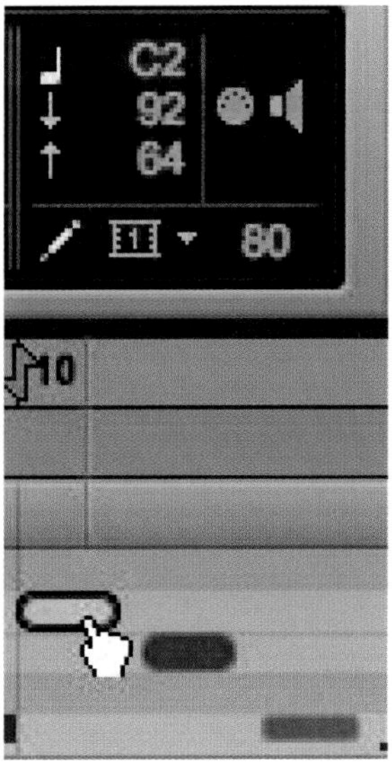

Figure 2.45 MIDI
Editing Controls
Area in the Toolbar
above a MIDI track
in the Edit window
with a selected
MIDI note, C2,
displayed in the
MIDI Editing
Controls Area
together with Its
On Velocity of 92
and Off Velocity
of 64.

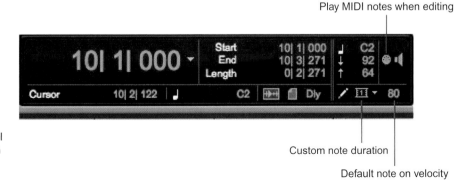

Play MIDI notes when editing

Custom note duration

Default note on velocity

Figure 2.46 MIDI
Editing Controls in
the Edit window
Toolbar.

The MIDI socket with loudspeaker icon to the right of the MIDI note parameters display can be used to enable or disable the Play MIDI Notes When Editing feature. When this is highlighted in green, MIDI notes will sound when you click on them while editing.

When you are manually inserting notes into a MIDI or Instrument track using the Pencil tool, the pop-up Custom Note Duration selector lets you choose the default note duration.

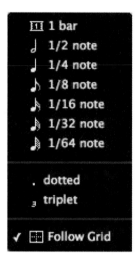

Figure 2.47
Custom Note
Duration pop-up
selector.

To the right of this, the Default Note On Velocity setting lets you define the default note-on velocity.

Cursor Values Display

Below the Counters, there is a display area that tracks the position of the cursor. This area also displays cursor values such as the MIDI note name when the cursor is moved vertically within a MIDI track or the Volume level in dBs when the cursor is moved vertically within a Master track.

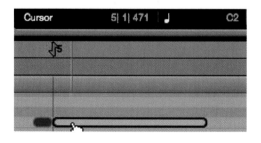

Figure 2.48
Cursor values
display.

Edit Window Indicators

Positioned directly below the Event Edit area, there are three small status indicators: Timeline Data Online Status, Session Data Online Status, and Automatic Delay Compensation.

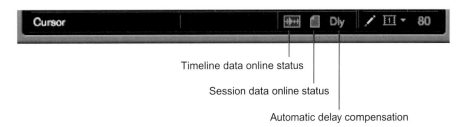

Timeline data online status

Session data online status

Automatic delay compensation

Figure 2.49 Edit Window Status Indicators.

If any files that are used in track playlists (i.e., visible in the Timeline: the main part of the Edit window) are off-line, being processed, or otherwise unavailable for playback, the Timeline Data Online Status indicator will turn red to warn you.

If any audio or fade files referenced by the session are off-line, being processed, or otherwise unavailable for playback, the Session Data Online Status indicator turns red to warn you.

When automatic Delay Compensation is enabled in the Options menu, this indicator shows "Dly" to indicate this.

Grid and Nudge

Separate displays and controls are provided for the Grid and Nudge features.

Figure 2.50 Grid and Nudge Displays and controls.

You can click on the small downwards pointing arrows to the right of these displays to open the pop-up selectors that let you change the Grid or Nudge values.

If you are working on music production, you would normally set these values to Bars:Beats, although Samples may be useful for certain types of edits, and Minutes:Seconds is useful if you are not working to Bars:Beats.

The Grid values are used in Grid mode to set the increments by which regions or MIDI notes can be moved forwards or backwards in the Edit window.

The Nudge values are used when "nudging" regions or notes forwards or backwards in the Edit window. (The word "nudge" used in this context means to move by a small amount.)

> **Tip**
>
> To increment or decrement the nudge values using the keyboard, hold the Command and Option keys (Mac) or Control and Alt keys (Windows) and press the plus (+) and minus (–) keys on your QWERTY or numeric keypads.

To nudge a region, regions, region groups, MIDI notes, or automation breakpoints, select these using the Time Grabber or Selector tool then press the Plus (+) and Minus (–) keys on the numeric keypad to nudge forwards and backwards in time. Any automation data relating to a selection of regions that you nudge will also be shifted.

> **Note**
>
> Pro Tools allows you to nudge material continuously in real time during playback. So, for example, you could nudge an audio region on one track to adjust the timing relationship between this and the rest of the tracks. Be aware, though, that there will be a brief delay before you actually hear the nudged version while playing back.

The Rulers

Immediately below the Edit window Toolbar you will find one or more "rulers." If you are recording music, the main ruler that you should be using will display Bar:Beats, or possibly Minutes:Seconds. At various times, you will also want to display Markers, Tempo, Meter, Key, and Chords rulers.

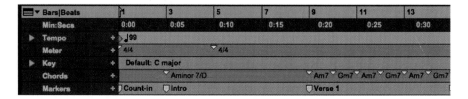

Figure 2.51
A Selection of Rulers showing below the Edit window Toolbar.

For certain types of edits, it can be useful to display the Samples ruler. If you are working for picture, you can have up to two Time Code rulers, and for film work you can display Feet+Frames.

You can choose which rulers to display either by selecting these from the Rulers submenu in the View menu or by using the pop-up selector at the far left of the main ruler in the Edit window.

Figure 2.52
Rulers pop-up
selector.

The Tempo ruler can be used to insert tempo changes anywhere in your music. Just click on the small "+" sign to open the Tempo Change dialog and insert a tempo change at the current location on the Timeline. The Tempo ruler can also be "opened" by clicking on the small arrow to the left of its name to reveal the Tempo Editor that allows you to edit tempo "events" graphically.

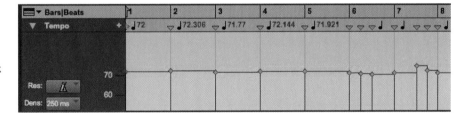

Figure 2.53
"Conductor" Track
Ruler for Tempo
with the Tempo
Editor revealed.

The Meter ruler lets you insert time signature markings at appropriate locations along the timeline. Just click on the small "+" sign to open the Meter Change dialog and insert a meter change at the current location on the Timeline.

The Key Signature ruler can be used to insert key changes anywhere in your music. Just click on the small "+" sign to open the Key Change dialog and insert a key change at the current location on the Timeline. Like the Tempo ruler, the Key Signature ruler can also be "opened" by clicking on the small arrow to the left of its name. This reveals the Key Signature Editor that allows you to edit key changes graphically.

Figure 2.54 Key Signature Editor.

The Markers ruler lets you insert markers at appropriate locations along the timeline. Just click on the small "+" sign to open the New Memory Location dialog and choose and name a marker and insert this at the current location on the Timeline.

The Chords ruler lets you add chord symbols to Pro Tools sessions in the Edit window (and in MIDI Editor and the Score Editor windows). Chord symbols in Pro Tools are simply markers that display chord symbols and guitar tablatures—they have no effect on MIDI data.

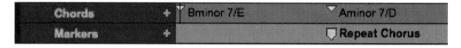

Figure 2.55 Rulers view showing the Chords ruler above the Markers ruler.

The Chords ruler lets you add, change, move, and delete chord symbols. To add a chord symbol, place the cursor in the Timeline where you want to add this. Click the Plus (+) button in the Chords ruler to open the Chord Change dialog at the current Timeline location. In the Chord Change dialog, you can select the name for the root of the chord, the chord type, the bass note of the chord, and the chord diagram that will be displayed in the Score Editor.

To change a chord symbol, double-click the Chord Symbol marker in the Chord Symbol ruler to open the Chord Change dialog again. If you want to reposition a chord symbol, you can just click and drag the Chord Symbol marker to any new location along the timeline. To delete a chord symbol, either Option-click (Mac) or Alt-click (Windows) on the Chord Symbol marker in the Chords ruler. Alternatively, you can make a selection in the Chords ruler that includes the chord symbols you want to delete and choose Clear from the Edit menu or just press Delete.

Tip

Another way to open the Chord Change, Key Change, Meter Change, Tempo Change, or New Memory Location dialogs that can be convenient is to move the cursor into the ruler while pressing the Control key (Mac) or Start key (Windows). The cursor changes to the Grabber with a "+," and when you click at the location where you want to place the Chord symbol, Key signature, Time signature, Tempo, or Marker, the appropriate dialog opens.

Edit Window Pop-up Selectors

Immediately beneath the rulers at the far left of the Edit window you will find two more pop-up selectors. The first of these duplicates the options available in the View menu for the Edit window, allowing you to show or hide Inserts, Sends, and so forth.

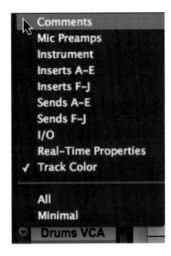

Figure 2.56 Edit window View options pop-up selector.

To the right of this, you will find the Linearity Display Mode pop-up. This lets you choose between Linear Tick (Bars:Beats) Display and a Linear Sample (absolute) Display.

Figure 2.57 Linearity Display Mode pop-up.

Track Heights

Pro Tools has seven preset track heights ranging from Micro to Extreme, plus the Fit to window command. MIDI and Instrument tracks also feature a single note view. Micro is the smallest possible Track Height, with six more choices ranging from Mini up to Extreme, plus "Fit To Window." "Fit To Window" fills the entire tracks area of the Edit window (at the current window size) with the selected track: use this setting when you need to focus on a single track for editing.

These track heights can be selected from the Track Options pop-up selector that appears when you click on the small arrow that can be found at the top left of each set of Track controls in the Edit window.

Figure 2.58 Track Options pop-up selector arrow.

Alternatively, you can press, click, and hold the mouse pointer onto the vertical rule to the left of the waveform display on any audio track to access the same pop-up selector, which can often be more convenient.

Figure 2.59 Click on the vertical rule to the left of the waveform display to reveal the Track height pop-up selector.

There is a mini keyboard at the left of each MIDI and Instrument track in the Edit window that lets you select and play notes on that track by clicking with the mouse pointer. To reveal the track height selector, you have to press and hold the Control key while pointing at and clicking on the mini keyboard at the left of MIDI or Instrument tracks—otherwise this action will select and play the notes on the track corresponding to the note on the mini keyboard that you touched.

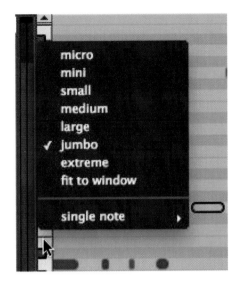

Figure 2.60 MIDI Track height pop-up selector— hold the Control key to access this.

Continuously Variable Track Heights

Pro Tools offers yet another way to change the height of a track, resizing in small increments, or continuously if you hold the Command key (Mac) or the Control key (Windows). Simply click and drag the lower boundary of the Track Controls column in the Edit window.

When you point your mouse at the bottom line of any Track Controls column (the area at the left of each track in the Edit window containing controls such as Mute, Solo, and so forth), the mouse pointer changes into a thick horizontal line with up and down arrows.

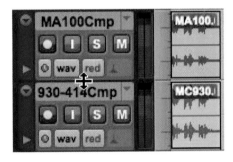

Figure 2.61 Mouse pointer poised, ready to adjust the track height.

If you then press and hold the mouse button and drag upwards or downwards, the track height will change—either in small increments or continuously, depending on whether or not you are also pressing and holding the Command key (Mac) or the Control key (Windows).

As usual with Pro Tools, there are short cuts provided to let you apply changes to all the tracks or to a group of tracks that you have selected. To continuously resize all tracks, Option-click (Mac) or Alt-click (Windows) and drag. To continuously resize all selected tracks, Option-Shift-click (Mac) or Alt-Shift-click (Windows) and drag.

Tip

The continuous track height resizing capability makes another couple of useful features possible:

To make all the tracks in which you have made an Edit selection just fit into the Edit window, press and hold the Command and Control keys Arrow (Mac) or the Control and Start keys (Windows), and then press either the Up or Down Arrow key (it doesn't matter which).

To make all the tracks that you currently have selected fit exactly into the Edit window, Option-Shift-click (Mac) or Alt-Shift-click (Windows) onto the Track Height pop-up menu and select Fit to Window.

Zooming Around

Getting used to the different ways to zoom in or out to see your audio and MIDI data displayed appropriately for the tasks you are approaching will take a little while. As usual, Pro Tools provides several ways to do the same things.

Veteran Pro Tools users will be familiar with the group of zoom arrow buttons near the top left of the Edit window that let you zoom audio and MIDI tracks horizontally or vertically. Click the left pointing or right pointing arrows to zoom all the track displays horizontally. Click the leftmost pair of upwards and downwards pointing arrows to zoom the audio waveform vertically, and use the rightmost pair of upwards or downwards pointing arrows to zoom the MIDI display vertically.

Tip

In recent versions of Pro Tools, you can simply point, click, and drag on any of these arrow-shaped zoom buttons to change the zoom levels continuously. Dragging to left or right zooms horizontally in a continuous fashion. Dragging up or down on the Audio Zoom In or Out button continuously zooms all the audio tracks vertically—while preserving any waveform height offsets.

There are five small buttons below these zoom buttons that are actually memory locations into which you can store different zoom levels of your choice. Just set the zoom levels you want using the arrow keys (or any other method), and then Command-click (Mac) or Control-click (Windows) on any of the five Zoom Presets to store this zoom level.

You can also use the Zoomer tool (the one that looks like a magnifying glass). Just point and click or drag in any track to zoom in, expanding the waveform or MIDI display, or hold the Option (Mac) or Alt (Windows) key while you click or drag to zoom out, contracting the display.

Users of other popular software, such as Cubase, will probably feel more comfortable using the Marquee Zoom feature, which works as follows since Pro Tools version 7.4:

If you hold down the Control key (Mac) or Start key (Windows) while using the Zoomer tool and drag to the left or the right, the display will shrink or stretch horizontally—a bit like a concertina. All tracks are affected equally using this method.

If you drag up or down within any track, the height of the display within the track will zoom vertically. Only the track that you are working with is affected using this method.

Holding the Shift key as well applies this to all tracks: so if you Control-Shift-drag (Mac) or Start-Shift-drag (Windows) in an audio track with the Zoomer tool selected, this will apply to all the audio tracks (or to all the MIDI or Instrument tracks if you drag in one of these).

You can set all the audio track waveform heights to match the waveform height of the topmost audio track in the Edit window by Command-Shift-clicking (Mac) or Control-Shift-clicking (Windows) on the audio waveform vertical Zoom button.

> ## Tip
>
> If you double-click the Zoomer tool, the waveform display zooms all the tracks horizontally until everything in your session just fills the Edit window. This is a very convenient way to reset the display so that it shows your whole session in the Edit window. This action also resets the waveform display within each track to its default height.

Pro Tools also offers a keyboard command that zooms the selection to fill the Edit window. Make an Edit selection first, and then press Option-F (Mac) or Alt-F (Windows) and *all* the MIDI and Instrument tracks will zoom horizontally until the current edit selection just fills the Edit window. *Selected* tracks will also zoom vertically to make sure that all the MIDI notes within the selection are visible.

Very often, you will want to zoom all the tracks out so that you can see an overview of your session. If you press Option-A (Mac) or Alt-A (Windows), this will zoom all the tracks out all the way horizontally and vertically. MIDI and Instrument tracks will automatically zoom vertically to display all notes.

> **Tip**
>
> The only way to become familiar with the way the zoom features work is to keep on trying them out on a regular basis until they become "second nature" to you—so go back and try these out now!

Zoom Toggle

Using the Zoom Toggle feature, you can make the Edit window behave the way you want it to when you change the Edit selection or Track selection.

The Zoom Toggle button is located just below the Zoomer tool. Just click on this to activate the Zoom Toggle feature, or, when the Command Focus is active, just press the "e" key on your computer keyboard.

To configure the way this feature works, you need to set various preferences and zoom behaviors in the Zoom Toggle section of the Editing Preferences window, which you can access from the Setup menu.

Here, you can set the way that both the Vertical and Horizontal Zoom will switch (i.e., toggle) when you enable the Zoom Toggle feature, and you can also set the way that the Track Height and Track View behave.

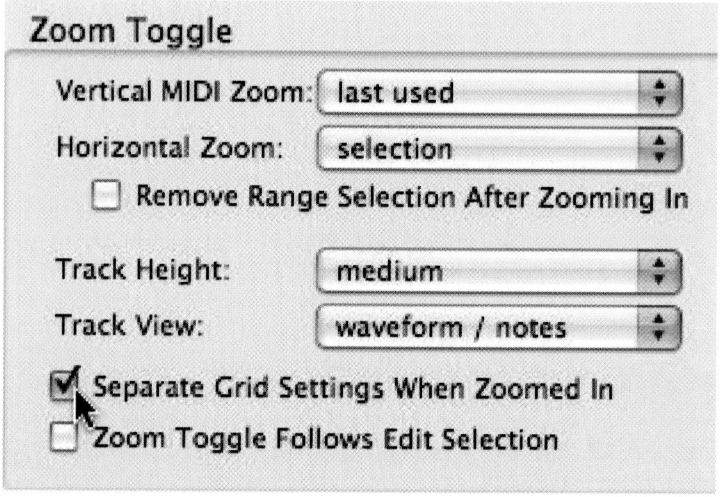

Figure 2.62
Zoom Toggle
Preferences.

The Vertical Zoom defaults to zooming to the current Edit Selection, with "Selection" chosen in the Zoom Toggle Preferences. If you are editing MIDI notes and you are in Notes view, and if you set the Vertical Zoom Toggle Preference to "Last Used," hitting the Zoom Toggle button zooms vertically to the last stored Zoom Toggle state for MIDI notes.

If you choose "Selection" as the Horizontal Zoom preference, then using the Zoom Toggle feature will cause the Edit window to zoom horizontally to the current selection, which is especially useful for audio editing. If you select "Last Used" instead, Zoom Toggle makes the Edit window zoom horizontally to the last stored Zoom Toggle state. This option is especially useful for editing MIDI notes.

A check box in the Zoom Toggle Preferences lets you "Remove Range Selection After Zooming In," in which case, the current Edit selection collapses and becomes an insertion point when you use the Zoom Toggle. This is useful when you know that you don't want to keep any selection you have made earlier when you zoom in using the Zoom Toggle feature.

You can also set preferences for how the Track Height behaves when you use the Zoom Toggle. If you choose "Last Used," then using the Zoom Toggle changes all tracks containing an Edit Selection to the last used Track Height. If you choose "Medium," "Large," "Jumbo," or "Extreme," all tracks containing an Edit Selection change to the Medium, Large, Jumbo, or Extreme Track Height, and if you choose "Fit To Window," all tracks containing an Edit Selection will change their Track Height to a size that will allow them to fit into the Edit window.

Quite often, you will have the need to change the Track View during a Pro Tools session, so the Zoom Toggle also lets you do this. To make the Zoom Toggle change the Track View for audio tracks to Waveform view and for Instrument and MIDI tracks to Notes view, choose "Waveform/Notes" in the Zoom Toggle Preferences. If you want Zoom Toggle to change the Track View to the last used Track View that was stored with Zoom Toggle, choose "Last Used" and if you don't want any change to the Track View, select the "No Change" option. With Pro Tools version 7.4 and newer versions, the Track View options also include "Warp/Notes."

If you want to keep the grid settings constant when you use Zoom Toggle, tick the option for "Separate Grid Settings When Zoomed In." Otherwise, the grid setting stored with Zoom Toggle will be recalled when you use the Zoom Toggle feature.

There is also a checkbox for "Zoom Toggle Follows Edit Selection." When this is ticked, Zoom Toggle automatically follows the current edit selection. When disabled, changing the edit selection has no effect on the currently toggled-in track.

> **Tip**
>
> If you are editing MIDI, the most sensible Zoom Toggle preferences choices are to set Vertical Zoom to "Last Used," Horizontal Zoom to "Last Used," Track View to "Waveform/Notes," and Track Height to "Fit to Window."

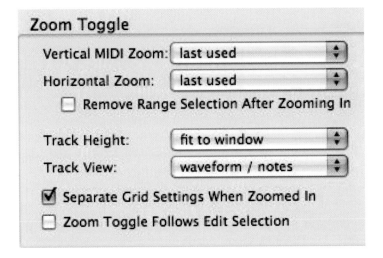

Figure 2.63
Recommended Zoom Toggle Preferences for MIDI.

> **Tip**
>
> If you are editing audio, the most sensible Zoom Toggle preferences choices are to set Vertical Zoom to "Selection," Horizontal Zoom to "Selection," Track Height to "Fit to Window," and Track View to "No Change."

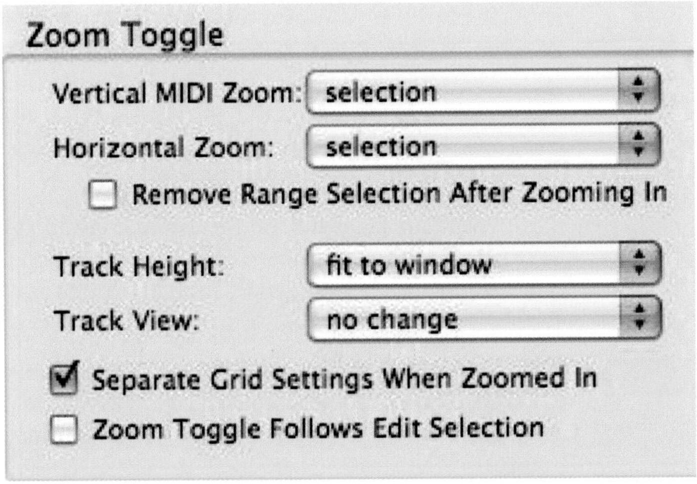

Figure 2.64
Recommended Zoom Toggle Preferences for Audio.

> ## Note
>
> When you have Zoom Toggle enabled and you select a smaller range of material, the window does not immediately zoom to encompass this new selection—it stays at the same zoom level. However, if you disable Zoom Toggle then enable it again, the window will zoom to encompass the current selection.

Scrolling Windows

There are various scrolling options in Pro Tools. The first of these, "None," is self-explanatory.

The "After Playback" scrolling option leaves the window where it is when you start playback and scrolls the view to the new position after you stop playback.

The "Page" scroll option keeps the display stationary until playback reaches the rightmost side of the page, and then it quickly changes the display to show the next section (i.e., page) of the waveform that will fit within the Edit window, and so forth.

There are two continuous scrolling options. The first of these, "Continuous," does exactly what it says—continuously moving the display during playback to keep up with the playback.

The second option, "Center Playhead," shows the "playhead"—a vertical blue bar indicating where the playback point is at in the Edit window—and moves the waveforms past this. The Blue Playhead line stays exactly where it is, in the center of the Edit window, and the items on the timeline move past this.

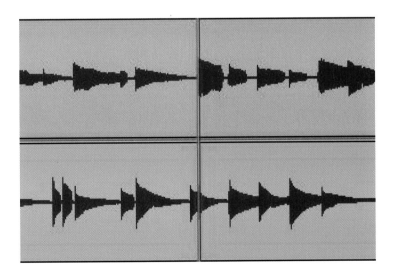

Figure 2.65
Continuous Scroll with Center Playhead.

Scrolling Behavior

With "Timeline Insertion Follows Playback" enabled in the Operation Preferences dialog, when you press Play, and then subsequently press Stop or Pause, and then press Play again, Pro Tools "picks up where it left off." In other words, when you stop playback, the insertion point "parks itself" at that point—instead of returning to the position it was at when you commenced playback.

If you are using the "Continuous" scroll option, the playback position jumps back to the position you started playing from when you pause or stop playback—unless you have enabled the "Timeline Insertion Follows Playback" option in the Operation Preferences dialog.

If you are using the "Center Playhead" scroll option, the playback position stays wherever it has reached at the moment you pause or stop playback. This happens because this scrolling mode overrides the "Timeline Insertion Follows Playback" preference, causing Pro Tools to behave as though this preference is enabled—even when it is not enabled.

Note that if "Timeline Insertion Follows Playback" was disabled before you entered the "Center Playhead" scrolling mode, it will still be disabled when you exit this mode, which is the way that most users prefer it to behave.

Playback Cursor Locator

The Playback Cursor Locator lets you locate the playback cursor when it is off-screen. For example, if scrolling is not active (set to "None"), when you stop playing back, the Playback Cursor will be positioned somewhere to the right, off the screen, if it has played past the location currently visible in the Edit window. Also, if you manually scroll the screen way off to the right, perhaps to check something visually, then the Playback Cursor will be positioned somewhere to the left off the screen.

To allow you to quickly navigate to wherever the Playback Cursor is positioned on-screen, you can use the Playback Cursor Locator button.

Note

The Playback Cursor Locator button only appears under certain conditions:

It will appear at the *right* edge of the Main Timebase Ruler after the playback cursor moves to any position *after* the location visible in the Edit window.

It will appear at the *left* edge of the Main Timebase Ruler if the playback cursor is located *before* the location visible in the Edit window.

Figure 2.66 The Playback Cursor Locator, the small blue arrowhead, can be seen at the top right in the Main Timebase Ruler.

A click on the Playback Cursor Locator immediately moves the Edit window's waveform display to the Playback Cursor's current on-screen location—saving you lots of time compared with any other way of finding this location.

> **Note**
>
> The Playback Cursor Locator changes color from blue to red when any of the tracks are record enabled.

Keyboard Focus

To allow the computer keyboard to be used to issue more commands than there are available keys, Pro Tools provides three types of commands: Keyboard Focus, the Region List Keyboard Focus, and the Group List Keyboard Focus.

Obviously, only one of these can be active at a time, so when you engage one Keyboard Focus it will disable the one earlier engaged.

Depending on which Keyboard Focus is enabled, you can use the keys on your computer's keyboard to select regions in the Region List, enable or disable Groups, or perform an edit or play command.

> **Tip**
>
> You can choose the Keyboard Focus by holding Command-Option (Mac) or Control-Alt (Windows) while you press 1 for the Commands, 2 for the Region List, or 3 for the Group List.

I normally have the Commands Keyboard Focus enabled so that I get fast access to all my favorite keyboard commands.

Commands Keyboard Focus

The Commands Keyboard Focus button for the Edit window is located in the upper right corner of the tracks pane in the Edit window. When you click on this, the letters "a" and "z," and the square outline, change to yellow to indicate that it is activated.

Figure 2.67
Commands
Keyboard Focus
button, "a–z":
located at the top
right of the tracks
pane in the Edit
window.

With the Commands Keyboard Focus activated, a wide range of single key shortcuts for editing and playback become available on your computer's keyboard. To see these commands listed, choose "Keyboard Shortcuts" from the Pro Tools Help menu. Note that there are some exceptions to this rule, e.g. with the Commands Keyboard Focus activated, you can simply press the 'B' key for Separate Region, Otherwise, use Command-E.

> **Tip**
>
> When the Commands Keyboard Focus is disabled, you can still use these keyboard shortcuts by pressing the Control key (Mac) or Start key (Windows) at the same time as the shortcut key.

Region List Keyboard Focus

When the Region List Keyboard Focus is selected by clicking the "a–z" button in the Regions List, audio regions, MIDI regions, and Region Groups can be located and selected in the Region List by typing the first few letters of the region's name.

Figure 2.68
Enabling the
Region List
Keyboard Focus.

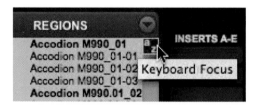

Group List Keyboard Focus

When the Group List Keyboard Focus is selected by clicking the "a–z" button in the Edit Groups list, you can enable or disable the Mix and Edit Groups by typing the Group ID letter (a, b, c, etc.) on your computer keyboard when using either the Mix or Edit window.

Figure 2.69
Enabling the
Group List
Keyboard Focus.

Channels and Tracks

In Pro Tools terminology, "tracks" are used to record audio (or MIDI) to your hard drive. These tracks appear in the Edit window and in the Mix window "channel" strips. Rather confusingly, what you might expect would be referred to as "channels" in the Mix window are referred to as Audio Tracks, Video Tracks, MIDI Tracks, Instrument Tracks, Auxiliary Inputs, VCA Master, and Master Faders in Pro Tools. Also, Pro Tools uses the term "channel" to refer to an actual physical input or output connection on whichever interface or interfaces you are using, so the documentation talks about a 96 I/O interface having 16 channels of audio input and output, for example. All this is straightforward enough to understand—but it is a little unusual when you are used to a tape recorder having tracks and a mixer having channels. It can be a little "jarring" mentally to talk about adjusting the mixer's "track" controls rather than the mixer's "channel" controls until you get used to the idea. But the documentation still talks about "channel" strips in the Pro Tools Mix window—each of which corresponds to a track in a Pro Tools session.

Voices and "Voiceable" Tracks

The concept of "voices" in Pro Tools needs to be thoroughly understood. The number of playback and recording "voices" is the number of unique simultaneous playback and record tracks on your system, a bit such as the concept of polyphony in a synthesizer, which dictates the maximum number of simultaneously playable notes. For example, with a Pro Tools HD 1 system, you can record up to 512 mono tracks (128 tracks in a basic Pro Tools 9 system)—but not all at once. And you can't play all these back at once either. You can only record and play back up to the limit of the available "voices" within your system, which depends on the type of card or cards you have in your system, and the sample rate you are working at. For example, a Pro Tools|HD 1 system can provide up to 96 voices of audio playback and recording at 44.1 or 48 kHz.

Avid spells out the rules for "voiceable" tracks as follows: The total number of "voiceable" tracks is the maximum number of audio tracks that can share the available voices on your system. (Mono tracks take up one voice. Stereo and multichannel tracks take up one voice per channel.) Voice limits are dependent on the session sample rate and the number of DSP chips dedicated to the system's Playback Engine. Pro Tools|HD systems can open sessions with up to 512 audio tracks, but any audio tracks beyond that system's voiceable track limit will be automatically set to *Voice Off*.

So, what happens when you run out of voices? The neat thing here is that you can assign more than one track to the same voice, and the track with the highest priority will always play back. For example, if a particular track contains an instrumental part that was only recorded in the verses of a song, and any empty audio regions are removed from the choruses, then you will hear any audio regions that exist on the next track (in order of priority) assigned to that voice play back in the chorus sections.

Just to make this even clearer, let me put it another way. When no regions are present within a particular time range on a track, the track with the next highest priority will be able to play back until a region appears again for playback in the first track—in which case the second track will stop playing back, and the first will resume playback.

The track priority depends on the order of the tracks in the Mix or Edit windows—with the leftmost track in the Mix window and the topmost track in the Edit windows having the greatest priority.

This system, referred to as "voice borrowing," makes Pro Tools behave as if it has many more tracks available than are allowed by the hardware. All you have to do is experiment with different combinations of track priorities, voice assignments, and arrangements of regions within your tracks to be able to play back many more tracks than you can with a conventional recorder.

On complex sessions with lots of tracks, you may still find yourself running out of voices at times. To help you manage your voice allocations, Pro Tools frees up a voice if you unassign the track's output and send assignments, or make the track inactive, or you can simply set the track's Voice Selector to "Off." Using the Voice Selector, a track's voice assignment can also be dynamically allocated or it can be explicitly assigned to a specific voice number. One restriction here is that Elastic Audio and RTAS plug-ins are not allowed on explicitly voiced tracks, so you must use Dynamically Allocated Voicing for these tracks instead.

> ### Note
> Only Pro Tools|HD systems allow tracks to be be explicitly assigned to a specific voice number.

Figure 2.70 Track Voice pop-up selector.

You can make tracks inactive by clicking on the track type icon that you will find to the right of the Voice Selector pop-up on each mixer channel strip and choosing "Make Inactive" from the pop-up selector that appears. You can also use this pop-up to make the track active if it has been deactivated. To give you visual feedback, mixer channels turn a darker shade of grey, and tracks in the Edit window are dimmed when inactive.

> **Tip**
>
> You can also use a keyboard command to toggle the track between active and inactive: Command-Control-click (Mac) or Control-Start-click (Windows) on the Track Type indicator in the Mix window.

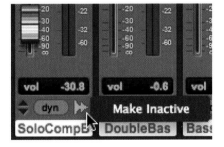

Figure 2.71 Clicking on the track type icon to access the Make Inactive pop-up.

Making some of your less important, or unused, tracks inactive is a great way of freeing up DSP resources and voices for use elsewhere, as all the plug-ins, sends, voices, and automation on inactive tracks are disabled. This feature also allows you to open Pro Tools sessions on systems with less DSP resources than were available when the sessions were created. Pro Tools will automatically make tracks inactive as necessary to allow sessions to be opened.

> **Note**
>
> Pro Tools|HD systems provide up to 160 Auxiliary Input tracks and a total of 256 internal mix busses. You can use up to 10 inserts and 10 sends per track (depending on the DSP capacity of your system). HD systems also support up to 128 Instrument tracks, 256 MIDI tracks, 128 VCA Master tracks, and multiple video tracks.

Mute Frees Assigned Voice

You can temporarily free up a voice by muting a track during playback, and if you have "Mute Frees Assigned Voice" enabled in the Options menu, the voice will be allocated to the next highest priority voiceable track that is assigned to the same voice.

This feature is only available for Pro Tools HD systems—standard Pro Tools systems do not allow voices to be specifically assigned.

> **Note**
>
> When you use the "Mute Frees Assigned Voice" feature, the computer introduces a delay of one or more seconds, depending on your CPU speed, before the mute or unmute instruction is carried out. If this bothers you, your only option is to turn it off. Also, muting a track using this feature will not make the voice available for use with QuickPunch, TrackPunch, or DestructivePunch recording. Larger DAE Playback Buffer Sizes can also increase the time between clicking the Mute button and the onset of muting.

Soloing Tracks

You can use the Solo buttons to solo any track or tracks at any time during playback, and if you solo a track that is a member of an active Mix Group, this will solo all other tracks that are a member of that active Mix Group as well.

The default Solo mode is called "Solo In Place" (SIP). In this mode, engaging a Solo button mutes all the other tracks so that the selected track can be auditioned independently. In SIP mode, Pro Tools lets you *solo safe* a track to prevent the track from being muted when you solo other tracks. To solo safe a track, Command-click (Mac) or Control-click (Windows) on the track's Solo button. The Solo button changes its appearance in Solo Safe mode, as can be seen in the accompanying screenshot.

Solo Safe is useful for tracks such as Auxiliary Inputs that are being used to sub-mix audio tracks, or for effects returns, to allow the audio or effects track to still be heard in the mix when other tracks are soloed. Without this feature, you would have to remember to also solo the appropriate Auxiliary tracks whenever you soloed a track, and this could be very awkward if the Auxiliary track is positioned a long way from the track you are soloing in the mixer.

> ## Tip
> It can also be useful to "solo safe" MIDI tracks so that their playback is not affected when you solo audio tracks.

Figure 2.72 Two stereo piano tracks bussed to an Auxiliary Input to form a submix with the Auxiliary Input in Solo Safe mode. Notice the way the solo-safed Solo button on the Aux track looks compared with the normal Solo button on the Double Bass track next to It, which Is muted because the piano tracks have been soloed.

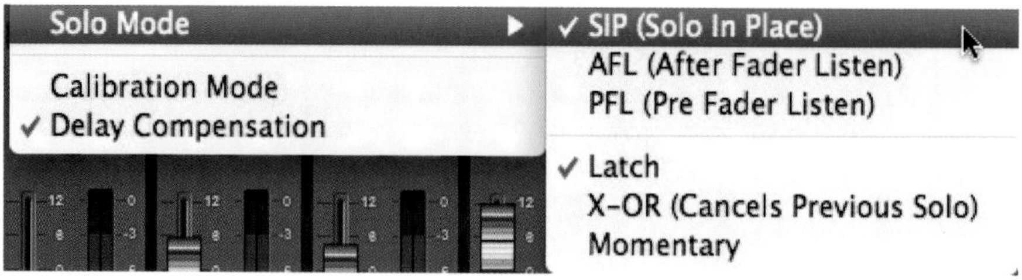

Figure 2.73
Choosing the Solo
mode from the
Options menu in
Pro Tools HD.

Solo Mode	▶	✓ SIP (Solo In Place)
		AFL (After Fader Listen)
Calibration Mode		PFL (Pre Fader Listen)
✓ Delay Compensation		
		✓ Latch
		X–OR (Cancels Previous Solo)
		Momentary

You can select two alternative Solo modes from the Options menu (Pro Tools HD only). Using these modes, After Fader Listen (AFL) and Pre Fader Listen (PFL), the Solo button can be used to route the selected track to a separate AFL or PFL output. You can configure the AFL/PFL output path in the Output page of the I/O Setup dialog.

> **Note**
>
> The Solo mode for all soloed tracks can be changed "on-the-fly" from any Solo mode to either SIP or AFL. Earlier soloed tracks will switch their solo behavior to the new mode. If you try switching the Solo mode for soloed tracks "on-the-fly" to PFL, all the earlier soloed tracks will be taken out of Solo before entering PFL mode to prevent any large boosts in level.

In AFL mode, the Solo button routes the track's *postfader/postpan* signal via the AFL/PFL output path. With AFL, the level you hear is dependent on the fader level for that track. In PFL mode, the Solo button routes the track's *prefader/prepan* signal to the AFL/PFL output path. With PFL, the fader level and pan are ignored, and the level you hear is dependent on the signal's recorded level.

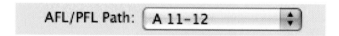

Figure 2.74 AFL/PFL Path Setting in the Output page of the I/O Setup dialog.

There is also a separate master level setting for AFL and PFL that affect the output of any or all tracks that you solo in AFL or PFL mode. When you are in AFL or PFL mode, you can set this level in the Mix or Edit windows. When you Command-click (Mac) or Control-click (Windows) on a Solo button on any track, a small fader appears that you can use to adjust the master level for the AFL/PFL Path.

> **Tip**
>
> To set the AFL/PFL Path level to 0 dB, Command-Control-click (Mac) or Control-Start-click (Windows) on any Solo button.

Figure 2.75
Setting the AFL/
PFL Solo level.

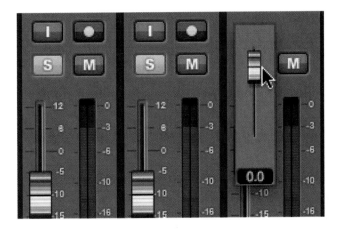

> **Note**
>
> You should set the AFL/PFL Output Path to "None" whenever you are not using this feature.
>
> Pro Tools allocates a substantial portion of its available DSP when using AFL/PFL mode, so setting the AFL/PFL Path to "None" will free up these DSP resources.

If you are not using a D-Control or D-Command work surface, where XMON is used to automatically switch the monitor source between the main output and the AFL/PFL output, you will need to choose suitable output paths both for your main monitoring and for your AFL/PFL monitoring in the I/O Setup and manually configure this so that the main Pro Tools output path is muted when you are using the AFL/PFL path. You can designate your main output pair as the outputs to be muted when tracks are soloed in AFL or PFL Solo mode using the AFL/PFL Mutes pop-up in the I/O Setup window.

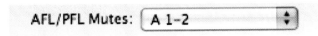

Figure 2.76 When tracks are soloed in AFL or PFL Solo mode, the output pair selected in the AFL/PFL Mutes pop-up in the I/O Setup window will be muted.

Solo buttons default to Latch mode, in which pressing other solo buttons also solos these tracks without disengaging earlier soloed tracks.

If you choose X–OR mode, pressing subsequent Solo buttons cancels earlier soloed tracks.

In Momentary mode (*Pro Tools HD and Pro Tools with Complete Production Toolkit 2 only*), a track is only in solo while its Solo button is being held down. To temporarily latch more than one solo button in Momentary mode, press and hold the Solo button on the first track that you want to solo, and then press additional Solo buttons. These will all remain soloed as long as one Solo button is held.

Finding Tracks in the Edit and Mix Windows

If you have lots of tracks in your Pro Tools session, there will be times when some of these are not visible on-screen because your computer monitor is not large enough. You can always use the scrollbars in the Mix or Edit windows to scroll along until you find the track you are interested in, but this can slow you down at times. However, if the track you are looking for is visible in either the Mix or the Edit window, you can instantly bring this track into view in the other window using the Scroll Into View command. You can access this command from the pop-up menu that appears when you Control-click (Mac) or Right-click (Windows) on the track name.

Figure 2.77
Control-click (Mac) or Right-click (Windows) on the Track Name, and You can choose "Scroll Into View" from the pop-up selector.

> **Note**
>
> The Split into Mono option is only available in this pop-up menu on multichannel tracks (stereo or surround formats).

Navigating to Tracks by Number

Pro Tools gives you the option of displaying a number for each track corresponding to its position in the Mix and Edit Windows. These numbers can be helpful when you want to know how many tracks you are working with, or how far along the Mix window you are, and especially if you want to find a track that is not currently showing in the window, and you can remember its number.

To activate this feature, make sure that Track Number is selected (with a tick next to its name) in the View menu. Then, you can navigate directly to any track using these Track Position Numbers. Simply open the Scroll to Track dialog from the Track menu, or by pressing Command-Option-F (Mac) or Control-Alt-F (Windows), and enter the Track Number for the track you want to view.

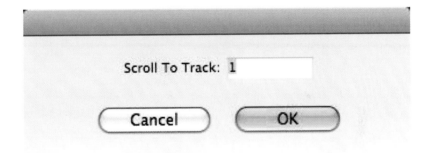

Figure 2.78
Scroll To Track dialog—accessible from the Track menu.

When you click OK, the track becomes selected, and the Edit window tracks scroll to bring the selected track as close to the top as possible. The Mix window tracks scroll to bring the selected track as close to the left as possible.

> **Note**
>
> If you change the positions of any tracks in the Mix or Edit windows, the Track Position Numbers will be automatically renumbered to keep them in numerical sequence.

Fit All Shown Tracks in the Edit Window

Sometimes you might want to fit all the tracks that you have showing in the Edit window into the visible part of the window, so Pro Tools provides a keyboard command to do this: Command-Option-Control-Up Arrow or Down Arrow (Mac) or Control-Alt-Start-Up Arrow or Down Arrow (Windows).

Obviously, there are limits to this—depending on how many tracks you have and how big you make the Edit window. All the tracks will be resized to an appropriate height to fit and fill the window, so if you just have one or two tracks, these might be resized to Jumbo height, but if you have lots of tracks and a small window size they might all go down to Micro size, as in the accompanying screenshot. And if the window is too small, and there are too many tracks, they still won't all fit. Nevertheless, this can be a very useful feature at times!

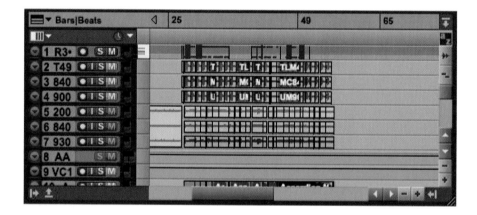

Figure 2.79
Fitting as many tracks as possible in the Edit window.

Note

If Zoom Toggle is enabled, fitting all tracks in the Edit window cancels Zoom Toggle.

"Do to All" and "Do to All Selected"

Here's a final tip for this chapter: two extremely useful keyboard commands that let you apply changes to all tracks or to whichever tracks you have selected. For example, you might want to set all the track outputs or all selected track outputs to the same destination.

To "Do to All," press and hold Option (Mac) or Alt (Windows) as you make your change. To "Do to All Selected," press and hold Option-Shift (Mac) or Alt-Shift (Windows) as you make your change.

Summary

Make sure that you know all the basic features of the user interface, and your patience will be rewarded when you realize how fast you can work with the system. Take time out to practice with features like Zoom Toggle and all the different ways that you can zoom the display. Zoom Toggle can deliver unexpected results at times, so you need to make sure that you have fully mastered this before using it on an important session.

Get used to the rulers (these are very easy to work with) and familiarize yourself with all the Toolbar controls and displays at the top of the Edit window. You will need to practice with the scrub and shuttle features if you have not used these before, but these should not present you with too many difficulties once you have spent some time shuttling around in the Edit window and scrubbing to look for edit points.

And make sure that you pick up as many keyboard commands as you can along the way! You can use the mouse to do most things, but it can be much, much quicker to press just one or two keys on the computer keyboard—or on the numeric keypad if you have one of these. Desktop computers normally have a numeric keypad to the right of the main keyboard, and you can buy separate numeric keypads for laptops or for any computer that doesn't have one of these.

Also, don't forget to keep your windows tidy and organized—hiding anything that you don't need to see, such as Region Lists, Inserts, Sends, unused tracks, or rulers that you are not using. A cluttered screen simply distracts—and you lose time that you can ill afford to lose on busy sessions if you have to look twice to find something. You should also make sure that you are clear about channels and tracks, about voices, and voice-able tracks, about how to mute and solo tracks, and how to find tracks in large sessions. Master the user interface and you will be well on your way to mastering Pro Tools!

In This Chapter

Time Operations, Tempo Operations, and Key Signatures

Introduction

One of the first things you ought to consider whenever you are working on music in Pro Tools is the tempo. If you are about to record a new piece of music, you will usually want to have the musicians play along to a click to help keep their timing accurate. This will also help to ensure that the music you record lines up with the bar lines in your Pro Tools session, so that you can edit using the grid and other editing features. If the musicians don't want to record to a click, there are ways to match the audio you record to the bars and beats in Pro Tools afterwards, but this will inevitably take more time to sort out.

You may also need to create tempo maps so that musicians can play faster or slower in different sections of the music along with meter changes, whenever these are needed. Even short pop songs sometimes have odd numbers of bars or beats, or changing meters, and these are very common in music-to-picture. The meter, or time signature, defines the number of beats in a bar and the way they can be subdivided, so you need to be familiar with all these things to get the best results.

It can also be very helpful to establish the correct key signature for your music, especially if you intend to use the Score Editor or key transposition features. And if you want to quickly check whether a guitar or keyboard or bass part or whatever will work better in another key, the Elastic Audio Pitch Transposition feature is the perfect tool.

Elastic Audio also lets you treat audio much like MIDI—quantizing and moving notes around more or less at will—using time compression and expansion algorithms in this case. You can use Elastic Audio to quickly beat match an entire song to the session tempo and Bar|Beat grid. Or you can quantize audio to a groove template so that other audio regions take on the same feel. For instance, you might take the feel from one drum loop that you like and apply that to another. And if you are a fan of Beatles-style recording effects, you can use the Varispeed algorithm to create tape-like effects with speed changes that cause pitch changes.

Beat Detective provides various kinds of analysis of tempo and "groove" for your audio and MIDI recordings. For example, you can use Beat Detective to establish a "groove" template that deviates from the standard grid so that you can match new recordings to the "groove" of a particular drum-machine, sequencer, or percussion loop. Beat Detective also lets you analyze the beats in any audio region, and then allows you to separate these so they can be moved around or edited for a variety of purposes. So you can conform beats to the grid in Pro Tools or make the session tempo follow the tempo that your drummer has established.

Tip

If you are learning Pro Tools for the first time, you should make the effort to get used to the terminology that Avid uses to describe the features of the system so that this all becomes very familiar to you as soon as possible. Then you will find everything less confusing; and when you need to look something up in the manual or remind yourself about how something works, you will be much faster.

Timebase and Conductor Rulers

In Pro Tools, you edit and arrange your music along the tracks in the Edit window by placing and moving the audio and MIDI regions along the timeline that runs from left to right along the window.

Regions and events can be anchored to specific points or locations along this timeline. You can view these locations as time in seconds and minutes; as SMPTE time code locations; as bar, beat, and tick locations; or in other formats, using the Timebase rulers that can be displayed along the top of the Edit window above the Conductor rulers.

The Timebase rulers can measure time in two different ways: sample-based time (absolute time) as shown in the Minutes:Seconds ruler and tick-based time (relative time) as shown in the Bars:Beats ruler. The relative time that the Bars:Beats ruler represents will change depending on the tempo and meter settings. Absolute time, as displayed in the Minutes:Seconds ruler, can never change, which of course is why it is called absolute time.

Any or all of the Timebase rulers can be displayed at the top of the Edit window in Bars:Beats; in Minutes:Seconds; as Time Code, which displays the Time Scale in SMPTE frames; or as Time Code 2, which lets you refer video frame rates that are different from the session Time Code rate or as Feet + Frames, which is used for film work. You can also display the Time Scale in samples, which can be very useful for precise editing tasks.

The most useful Timebase rulers for music production are Bars:Beats and Minutes:Seconds; so you should deselect the ones that you are not using to remove the potential for confusion and to conserve screen space.

> **Tip**
>
> The Timebase rulers may be used to define *Edit selections* for track material and *timeline selections* for record and play ranges. For example, using the Selector tool, you can drag along any Timebase ruler to select material across all the tracks in the Edit window. To include the Conductor tracks in the selection, just press Option (Mac) or Alt (Windows) while you drag. Often, you will want to link the Edit and timeline selections, but sometimes you will want to make different Edit and timeline selections. You can make or break the link between these using the Link timeline and Edit Selection button located just below the tools in the Edit window.

The Conductor rulers include the Tempo, Meter, Key Signature, Chords, and Markers rulers. Tempo and Meter events entered into these rulers will affect the timing of any tick-based tracks and also provide the tempo and meter map for the Bar:Beat grid and Click. Pro Tools allows you to edit Tempo events either in the Tempo ruler or by using the Tempo Editor as shown below.

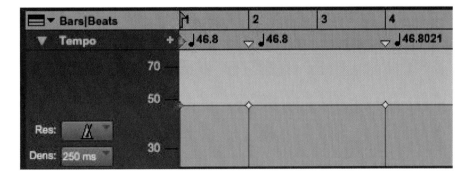

Figure 3.1
Tempo Editor.

You can enter key signatures and make key changes using the Key Signature ruler. This is useful when transposing MIDI notes, for example. When you choose "Transpose in Key" in the Event Operations Transpose window and enter a particular number of scale steps, Pro Tools works out what the root note of the key should be based on the key signature that applies at this point along the timeline and then transposes any selected MIDI notes by the number of steps that you specify. You can also change the key of audio regions.

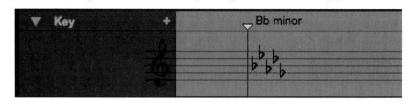

Figure 3.2
Key Signature
Editor.

The Chords ruler lets you display Chord symbols for musicians to follow and the Markers ruler lets you display markers to indicate which section of the music is which.

Figure 3.3
Chords and
Markers.

Markers are actually Memory locations that can be used for other purposes, but their most basic function, that of marking out the sections of your music, is probably their most important function. It is a good idea to define markers for any musical session in Pro Tools as early as possible.

Choosing the Marker Reference

When you create or edit Marker (or Selection) Memory Locations in the New or Edit Memory Location dialogs, you can choose whether they should have an Absolute (sample-based) or Bar:Beat (tick-based) reference using the Reference pop-up.

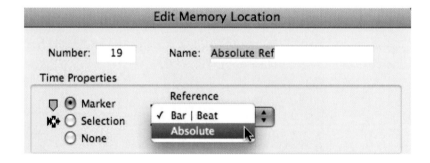

Figure 3.4
Choosing the
Marker Reference.

If you choose Bar|Beat, the Memory Location will be tick-based and its bar|beat location *will not* change if the tempo is changed. However, its sample location *will* change.

If you choose Absolute, the Memory Location will be sample-based and its bar|beat location *will* change if the tempo is changed. But its sample location *will not* change.

Bar|Beat Reference Markers appear as yellow chevrons in the Markers ruler and Absolute Reference Markers appear as yellow diamonds, so that you can distinguish between the two types of markers.

Figure 3.5
Bar|Beat Reference
and Absolute
Reference Markers.

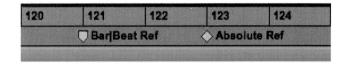

Setting the Time Scales

The Main Time Scale that you should choose for the session will normally be Bars:Beats for music production. This will be displayed as the Main Counter in the Transport window and at the top of the Edit window. This counter will also be used for the Start, End, and Length values next to the Main Counter; for the Grid and Nudge values in the Edit window; and for any Pre- and Post-roll amounts that you set.

You can also choose which Sub Counter Time Scale to display below each Main Counter. The obvious choice for music production would be Minutes:Seconds so that you always get some feedback about how long regions or complete sessions last.

> **Note**
>
> When the Main Time Scale is set to Bars:Beats and you are using tempo changes, you should set the Linearity Display Mode to Linear Tick Display. This will keep the Bars:Beats ruler fixed at the selected zoom level, and sample-based rulers such as Minutes:Seconds will scale to fit any tempo changes while bar lengths remain constant.

Meter, Tempo, and Click

If you intend to do anything but the most basic recording and playback of musical audio material, you will need to make sure that the bar and beat positions in the audio line up with the bar and beat positions in Pro Tools. It is absolutely essential that the bars correspond correctly if you want to be able to navigate to bar positions and use Grid mode and other features to edit your audio with ease. If you are recording musicians, you can generate a click for the musicians to play along. If they do this well, then the music they play will be in time with the click and will line up with the bar lines in Pro Tools.

Audio that is recorded without listening to a click can still be aligned to bar and beat boundaries in Pro Tools either by using Beat Detective or by using the Identify Beat command, but this can take some time. You can also record MIDI with or without a click and either manually add Bar|Beat Markers or generate a tempo and meter map using Beat Detective.

Creating a Click Track

In older versions of Pro Tools, the way to create a click track was to create a mono Auxiliary Input and insert the Avid Click plug-in or a virtual instrument plug-in or an external MIDI device to play the click. You can still do any of these things if you like, but in more recent versions, you can use the Create Click Track command, which you will find in the Track menu. When you choose this command, Pro Tools create a new Auxiliary Input track named "Click" with the Click plug-in already inserted and with the click track's Track Height conveniently set to Mini in the Edit window. To hear the click, you also need to highlight the Metronome icon in the Transport window by clicking this, or by pressing the "7" key on your numeric keypad, or by selecting "Click" in the Options menu.

Figure 3.6
The Click Plug-in.

Tip

To avoid cluttering up your Mix and Edit windows in Pro Tools, you can hide the click track using the "hide" feature in the Tracks List. If you solo-safe your click track (using a Command-Click [Mac] or Control-Click [Windows] on the track's solo button), you can hide your click track in the tracks list and use the "7" key on your numeric key pad to turn the click on and off at any time. Because the track is "solo safed," you will always hear the click when other tracks in your session are soloed.

You can use the Click/Countoff Options dialog, available from the Setup menu, to choose when the click should be active and to set the MIDI notes, velocities, and durations that should play if you are using a virtual instrument or an external MIDI device to produce the click sound. The MIDI parameters in the Click/Countoff Options dialog do not affect the Click plug-in—this is controlled internally by the Pro Tools software. You can also set the number of count-in bars that you prefer and whether you want to hear these sounds every time you playback or only when recording.

> **Note**
>
> If you are using the internal Click plug-in, leave the Output popup set to "None." If you are using an external device or a software synthesizer to produce your click sound, you can choose this device or synthesizer using the Output popup.

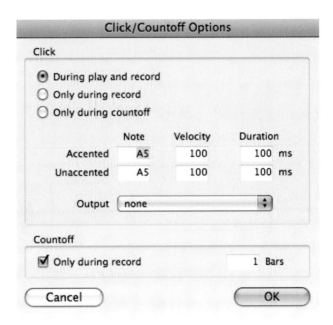

Figure 3.7
Click/Countoff
Options dialog.

Time Signatures and Clicks

When you are setting up a click for musicians to play along to, you may need to change the time signature if the music is not in the default 4/4 meter. This will be very obvious if the music is in time signatures 3/4 or 5/4 which are used in popular music. But it may not be quite as obvious if the music is in 6/8 or 12/8.

And if the music is in 6/8, for instance, you may find it better to use a dotted quarter note value for the click. This way, you would get two clicks in each bar. Otherwise, if you use an eighth-note click, this would give you six clicks in each bar, which can sometimes be distracting for the musicians.

To set the meter, double-click on the Current Meter display in the MIDI Controls to open the Change Meter dialog.

Figure 3.8
Double-click the
Current Meter
display in the MIDI
Controls to open
the Change Meter
dialog.

In this dialog window, make sure that you are at the correct location in your session by typing in the correct bar and beat, and tick values if these are not already correct. If the music uses the same time signature throughout, then you should enter this meter value at the start of your song or music composition. Obviously, if the meter changes anywhere in the song, you will also need to enter meter changes at these locations.

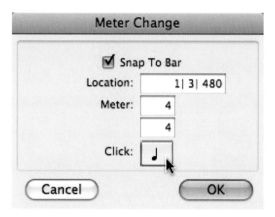

Figure 3.9
Changing the
Click resolution.

The Meter Change dialog window has a popup selector that lets you change the click "resolution."

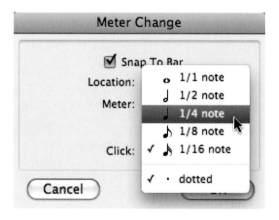

Figure 3.10
Choosing dotted
1/4 note click
resolution.

Here you can choose the note subdivision that makes most sense to use. For example, you might choose a dotted 1/4 note resolution when you are in 6/8 meter so that you get just two clicks per bar.

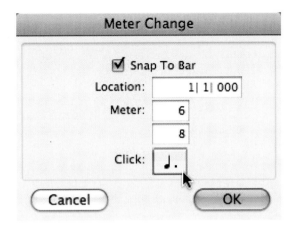

Figure 3.11
Meter set to 6/8
from bar 1, beat 1,
with the click
resolution set to
a dotted quarter
note.

Tempo

Tempo is one of the most fundamental aspects of music. After all, if there is no tempo, the music goes nowhere! And if the tempo becomes erratic, the music is not "working" properly. Also, when you are making a recording, getting the tempo to feel "right" can make all the difference between a successful recording and the one that does not inspire people to listen to it. Of course this can be a subjective matter that not everyone will agree about, but most musicians will perform much better on a particular piece of music if the tempo makes them feel "comfortable." And this good feeling will usually communicate itself to the listeners. So it is very important to pay attention to tempo!

Figure 3.12
Setting the tempo
using the MIDI
tempo controls.

Setting Tempo Manually

You can set a tempo for your session manually using the MIDI tempo controls in the expanded Transport window. When the Conductor icon is not selected, you can either type the tempo or drag the mouse pointer in the tempo field to change the tempo.

Tapping the Tempo

If you are trying to match the tempo of the session to the tempo of a recording of some music and you have no idea what this should be, just click on the numeric tempo field to highlight it, as in the screenshot (Figure 3.12), and tap on the "t" key on your computer keyboard. If you can tap accurately in time with the music, Pro Tools will work out the tempo from your taps and display this for you. Even if you don't get the tempo exactly right, you should at least be able to get this "into the right ballpark" very quickly using this method.

Using Tempo Events

When you have decided on a tempo for your session, it makes good sense to enable the Conductor track by selecting the Conductor icon in the Transport window and enter the correct tempo for your session into the Tempo ruler's Song Start Marker.

> ### Note
>
> The small, red Song Start Marker located at the beginning of the Tempo ruler defines the initial tempo of the Session and defaults to a tempo of 120 BPM. You can move the Song Start Marker by dragging it to the left or right in the Tempo ruler, but it cannot be deleted.

To edit the default tempo, go to the Tempo ruler and double-click the red Song Start Marker triangle at the start of the track to open the Tempo Change dialog.

Figure 3.13
Clicking on the Song Start Marker in the Tempo Ruler.

Type the tempo you want to use for your session in the BPM field and click OK to replace the default 120 BPM value at the start of the song.

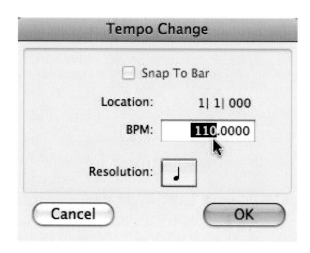

Figure 3.14
Tempo Change dialog.

Building a Tempo Map

If you know what tempos you want to use for your music, you can build a tempo map for the Conductor track by inserting Tempo events into the Tempo ruler at the locations where you want the tempo to change.

To insert Tempo events, click in the Tempo ruler at the location where you want to insert the tempo event, and then click the Add Tempo Change button to the left of the Tempo ruler to open the Tempo Change dialog.

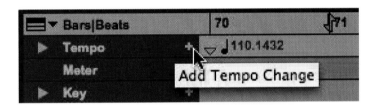

Figure 3.15
Adding a Tempo
Change to the
Tempo Ruler.

In the Tempo Change dialog, you can type the exact location where you wish to add the new tempo (very useful if you just clicked somewhere nearby this to open the dialog), type the new tempo in BPM, and click OK.

> **Tip**
>
> To save typing the location if you are close to this, you can tick the "Snap To Bar" option to place the new tempo event exactly on the first beat of the nearest bar.

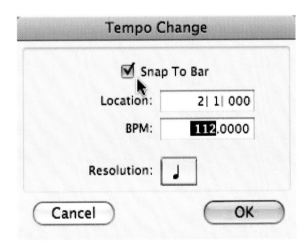

Figure 3.16
Tempo Change
dialog with Snap
To Bar option
selected.

Editing Tempo Events

If you need to adjust the position of your Tempo events, you can simply drag these back and forth along the timeline. When you drag a Tempo event to a new location, the sample and SMPTE locations for the event are also updated, and any neighboring MIDI events and audio regions on tick-based tracks, along with the ruler, shrink or expand as necessary to adjust for the new tempo location. However, the BPM value for the dragged tempo event remains constant, as do any other Tempo events in the session.

To delete a Tempo event, you can simply Option-click (Mac) or Alt-click (Windows) on the Tempo event. To delete several Tempo events at once, drag to select these using the Selector tool and then press the Backspace key or choose Clear from the Edit menu.

If you double-click with the Selector tool in the Tempo ruler, all the Tempo events will get selected. You can then cut, copy, and paste these using the standard computer keyboard commands.

Tempo Operations

You can also apply tempo changes to a time selection using the Tempo Operations dialog, which you will find in the Event menu's Tempo sub-menu. There are six Tempo Operations windows to choose from here, including Constant, Linear, Parabolic, S-Curve, Scale, and Stretch.

The first of these, which you will probably use the most often, lets you set a constant tempo at a specified bar position or between two specified bar positions. The next three windows let you create a linear, parabolic, or S-curved "ramp" of tempo changes between a pair of specified bar positions. The final two windows let you scale or stretch existing tempos between the specified bar positions. Each window has a basic set of controls and a more advanced set that is revealed when you check the "Advanced" check box.

For instance, to change the tempo between specific locations (which returns to the original tempo afterwards), you would use the Constant Tempo Operations dialog. To see how this works, select, say, Bar 5 to Bar 7 by dragging the cursor along the rulers, and then open the Constant Tempo Operations window from the Event menu. Type the new tempo into the Tempo field and tick the box "Preserve tempo after selection." This causes the tempo to go back to the original value after selection of the new tempo.

To ramp up the tempo from one value to another using a series of equal increments over a range of bars, you can use the Linear Tempo Operations dialog to specify the selection of bars and beats, the start tempo, and the end tempo. Using the Advanced settings, you can also specify the resolution and density of the Tempo events that you are creating across this range of bars.

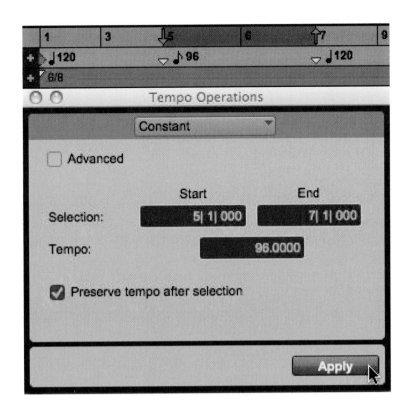

Figure 3.17
Applying a
constant Tempo
change to a range
of bars.

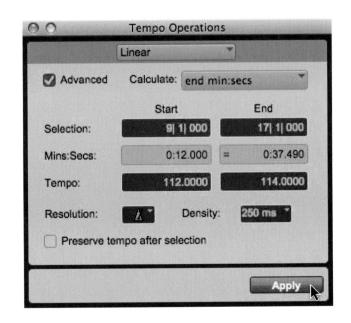

Figure 3.18
Using the Linear
Tempo Operations
dialog to ramp
between tempos
over a range
of bars.

Graphic Tempo Editor

Pro Tools often provides various alternative ways of doing things and tempo editing is no exception! If you click on the small arrowhead at the left of the Tempo ruler, it will expand downwards to reveal the Tempo Editor. Here, you can see tempo information represented graphically.

> **Note**
>
> Tempo events are visible in the Graphic Tempo Editor only when the Conductor is enabled in the Transport window.

You can change the location or value of any Tempo events by dragging these to left or right, or up or down.

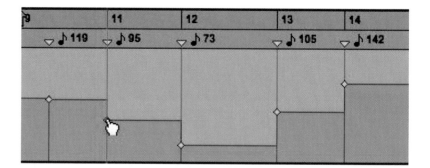

Figure 3.19
The Graphic
Tempo Editor.

Selected Tempo events can be copied and pasted, nudged, or shifted; and you can raise or lower any group of selected Tempo events using the Trimmer tool. First, make a selection that encompasses the Tempo events you want to trim, and then just click and drag up or down using the Trimmer tool.

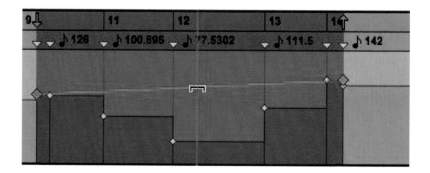

Figure 3.20
Using the Trimmer
Tool in the Graphic
Tempo Editor.

You can also "draw" in the new Tempo events, replacing existing ones, using the Pencil tool. The Line tool lets you draw straight lines, and with this tool, the Tempo values change in steps according to both the Tempo Edit Density and Resolution. There are seven drawing 'shapes' available for the Pencil tool. These can be selected by clicking and holding the mouse button down on the Pencil tool icon, moving the mouse to the required tool shape withe the button held, then letting go of the mouse button when the required tool shape is selected. The Free Hand tool lets you draw freely by moving the mouse, producing a series of steps that depend on the Tempo Edit Density setting. The Parabolic and the S-Curve tools let you draw the best possible curve to fit your freehand drawing, again producing a series of steps that depend on the Tempo Edit Density setting. The Triangle, Square, and Random Pencil tool shapes cannot be used to create Tempo events.

There are two popup selectors located at the left of the Tempo Editor that let you choose Tempo Resolution and Tempo Edit Density settings for the Tempo events that you draw. Using the "Res" popup selector, you can choose a Tempo Resolution or BPM rate by selecting a beat note value. The beat note value that you select will normally be based on the meter. For example, in 4/4, the beat would be a quarter note; and in 6/8, the beat would usually be a dotted-quarter note.

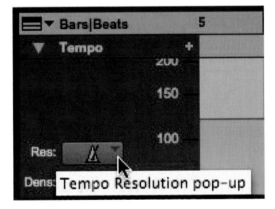

Figure 3.21
Tempo Resolution pop-up selector, located above the Tempo Density pop-up.

If you select Follow Metronome Click, Tempo events created by drawing with the Pencil tool will automatically mirror the click values set by Meter events in the Meter ruler.

> **Note**
>
> Although a tempo curve can have different BPM values if there are meter click changes within the selected range, it can be a complicated business to set Meter events separately for each Tempo event; so Avid recommends selecting Follow Metronome Click in most cases.

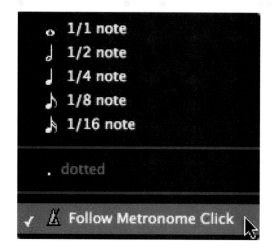

Figure 3.22
Tempo Resolution
selector.

The Tempo Editor also lets you specify the density of Tempo events created in the Tempo ruler when you draw a tempo curve with the Pencil tool. If you click on the Tempo Edit Density (Dens) selector, you can use the pop-up menu to select a suitable time value either as a note sub-division or in milliseconds.

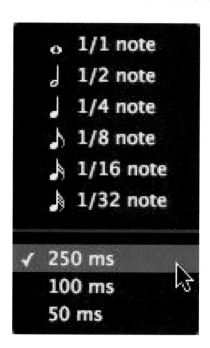

Figure 3.23
Tempo Edit
Density selector.

Changing the Track Timebase

By default, Pro Tools uses sample-based time for audio, measured in minutes and seconds, and tick-based time for MIDI, measured in bars and beats.

However, it is possible to change the Timebase of audio tracks to be tick-based or MIDI tracks to be sample-based.

If the track's height is set to medium or large, the Timebase selector button, with its tick-based or sampled-based icon, will be visible among the track controls. When you click on this, a popup selector appears, allowing you to change the Timebase for the track.

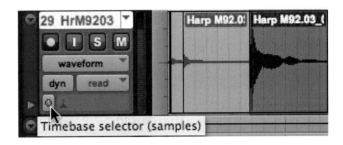

Figure 3.24
Timebase Selector.

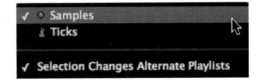

Figure 3.25
Popup Timebase
selector.

If the track's height is set to small or less, you can click the small arrow to the left of the Track Name to access the Track Options popup menu. Here you can choose the timebase using the Track Timebase submenu.

Figure 3.26
Track Options
arrow.

Figure 3.27
Track Options
Timebase
Sub-menu.

By default, when you change a track's timebase, the change is applied to any alternate playlists that exist on the track. If you want the change to be applied only to the active playlist, simply deselect the "Selection Changes Alternate Playlists" option.

> **Note**
>
> When you change the timebase for an audio track that is part of an active group, all the tracks in the group will change to the same timebase.

Sample-Based Audio and MIDI

Audio is stored as individual audio samples within audio files on disk. In Pro Tools, regions representing all or parts of these files are used to represent the audio on tracks in the Edit window. Audio tracks are designated by default as sample-based, so any particular audible sound is located at a particular sample location on the timeline. These locations are absolute locations in time, measured in samples from the start of the session.

If you change the tempo of the session, the bars and beats will move relative to the absolute (unchanging) locations of the sample-based audio regions, but these audio regions and the audio that you hear playing back will not be affected in any way.

> **Note**
>
> If you make a MIDI track sample-based, all the MIDI events in the track will have an absolute location on the timeline—just like sample-based audio regions—and these will stay fixed to this time location, no matter what tempo or meter changes take place.

Tick-Based Audio and MIDI

MIDI (and Instrument) tracks use tick-based timing, which tie events to bar|beat locations—not to sample locations. Tick-based locations are relative locations in time measured in bars, beats, and ticks from the start of the session. So, if you change the tempo of the session, a MIDI region located at a particular bar|beat location will not move—but the location of any sample-based regions in the same session *will* change relative to this bar|beat location.

If you increase the tempo on a tick-based track, the data plays back more quickly, and so the individual events take place sooner in time relative to any sample-based audio in the session. If you decrease the tempo, events will play back more slowly, and so the individual events will take place later in time relative to any sample-based audio in the session. Another way of thinking about this is to keep in mind that the bars and beats play back more quickly and last for a shorter time at faster tempos and vice versa.

If you are using Elastic Audio, REX, or ACID audio files; you will want to change the sample-based audio tracks to be tick-based, so that the tempo of the audio will automatically change when you change the tempo of the session.

> **Note**
>
> Pro Tools Elastic Audio uses exceptionally high-quality transient detection algorithms, beat and tempo analysis, and Time Compression and Expansion processing algorithms. Tick-based Elastic Audio tracks actually change the location of samples according to the changes in tempo, because the audio stretches or compresses to match these tempo changes.

Keep in mind that tempo changes will affect the locations of the start points (or the sync points if these are set differently) of any standard (i.e., non-Elastic Audio) audio regions in tick-based audio tracks. You can slice a drum loop into beats in Pro Tools, make the track tick-based, and it will play each slice back earlier or later according to the tempo—in a similar way to REX files.

> **Note**
>
> In tick-based audio tracks, the location of an audio region is determined by the region's start point, unless the region contains a sync point. If the region contains a sync point, the sync point determines where the audio region is fixed to the grid.

Time Scales and Tick Resolution

When you are working on music in Pro Tools, you will normally set the Main Time Scale to Bars:Beats with the Sub Counter Time Scale set to Mins:Secs. Because you are working with music, it will make most sense to use bar, beat, and tick values for a number of operations including placing and spotting regions, setting lengths for regions or MIDI notes, locating and setting play and record ranges (including pre- and post-roll), specifying settings in the Quantize and Change Duration pages of the Event Operations window, and setting the Grid and Nudge values. You would not normally want to set these using samples, although this may be useful at times.

With the tick-based Time Scale, the bars and beats are sub-divided into clock "ticks" with each quarter note represented by 960 ticks. This is often quoted as a tick resolution of 960 PPQN (pulses per quarter note), and all MIDI sequencers have a similar specification. Some MIDI sequencers have the same resolution as Pro Tools, while others offer less or more PPQN, and some allow you to select the resolution that you feel is appropriate.

> **Note**
>
> In Pro Tools, the actual internal MIDI resolution is 960,000 PPQN, with 960 PPQN used as the display resolution so that the numbers are more manageable.

The Linearity Display Mode

You can choose to view the timeline using either a Linear Sample scale or a Linear Tick scale that you can select using the Linearity Display Mode pop-up menu. The Linear Sample scale uses an absolute timebase while the Linear Tick Scale uses a relative timebase; so MIDI and Instrument tracks, audio tracks, and Tempo curves will look and behave quite differently depending on which of these you choose.

Figure 3.28
Linearity Display
Mode Selector.

When you are working primarily with sample-based material, which is usually the case when recording or mixing audio, you should use the Linear Sample Display mode—especially if you intend to align the tempo and meter map to sample-based events. In Linear Sample Display mode, because the timeline display is sample-based, the locations of tick-based events such as bars and beats will shift whenever you change tempo. So, for example, the locations of the bars and beats will move to different time positions—as you will see if you display minutes and seconds on the sample-based timeline.

While using the Linear Sample display, the Min:Secs ruler shows each second occupying the same distance along the Timeline. In the screenshots shown here, you can see that the number of seconds displayed stays the same (with no change made to the zoom level), while the corresponding number of beats displayed depends on the tempo. At slower tempos, fewer bars and beats play back in a given time while at faster tempos more bars and beats play back in that same time interval. Another way of looking at this is to notice that, as the tempo increases, the length of the bars along the timeline gets shorter.

Figure 3.29 Each beat at 93 beats per minute occupies around 2.6 seconds. So in 4 seconds shown in this example, less than 2 complete bars will play back.

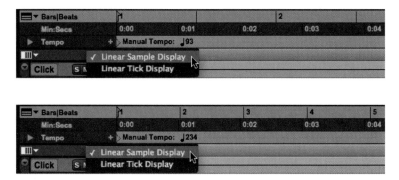

Figure 3.30 Each beat at 234 beats per minute occupies about 1 second. So with the Linear Sample display selected, you will see more bars playing back within the same number of seconds along the timeline—4 complete bars playing back in the 4 seconds displayed in this example.

When you are working with a tick-based material, such as MIDI notes or Elastic Audio events, you may find it more useful to switch to the Linear Tick Display mode. When you make tempo changes in Linear Tick Display mode, the bars and beats will remain fixed in the timeline, while sample-based events will move to different bar and beat locations along the timeline. In Linear Tick Display mode, all the bars are evenly spaced, whatever is the tempo. So, for example, sample-based audio regions will expand along the timeline as you increase the tempo—although they will still play back the audio that they contain in the same length of time. Tick-based material, such as MIDI notes or Elastic Audio material on tick-based tracks, will play back at whatever tempo you set the session to.

Figure 3.31 At 93 BPM, about one and a half bars are displayed here in Linear Tick mode, while the time elapsed shows 4 seconds.

Figure 3.32 At 234 BPM, one and a half bars are still displayed with the timeline display in Linear Tick mode (with no change made to the zoom level), but the time elapsed shows one and a half seconds.

Tip

You should use Linear Tick Display when drawing tempo changes, for example, because using the Linear Sample Display causes Bar|Beat-based material to move, making it awkward to work on tick-based material such as Tempo events.

Note

When the Tempo Edit Density setting (in the Tempo Editor) and the Linearity Display Mode setting are both set to either a Bars|Beats Time Scale or to an Absolute Time Scale, Tempo edits appear evenly spaced. When one of these is set to an absolute Time Scale and the other to a Bars|Beats Time Scale, the number of Tempo edits will appear to increase or decrease over time (against absolute time).

Tempo Events and Bar|Beat Markers

Tempo events are intended for use with tick-based tracks, while Bar|Beat Markers are for use with sample-based tracks. They cannot be mixed; therefore if a session contains Tempo events and you insert Bar|Beat Markers, the existing Tempo events are converted to Bar|Beat Markers (and vice versa).

> **Note**
>
> Tempo events are displayed as small green triangles in the Tempo ruler. Bar|Beat Markers look similar to Tempo events, but are blue in color.

Tempo events can be manually converted to Bar|Beat Markers and vice versa using the Tempo Ruler popup menu. When you hold down the Command (Mac) or Control (Windows) key on your computer keyboard and click the "Add Tempo Change" button, the Tempo ruler popup menu opens, where you can select which of these you wish to use.

Figure 3.33
Clicking while holding the Command (Mac) or Control (Windows) key to convert Tempo events to Bar|Beat Markers.

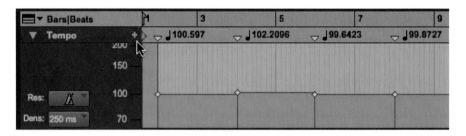

Figure 3.34
Using the Tempo Ruler popup menu to change Tempo events to Bar|Beat Markers.

> **Note**
>
> If you build a tempo map in a MIDI sequencer such as Digital Performer, Logic, or Cubase and then transfer this to Pro Tools using a MIDI file, this tempo map will appear as Tempo events in Pro Tools.

If you are working primarily with MIDI instruments in Pro Tools, it makes sense to use Tempo events for any tempo changes. On the other hand, if you are working primarily with audio and you want to map out where the barlines and beats fall within the audio material that was not recorded to a click, it makes better sense to use Bar|Beat Markers.

Building a Tempo Map by Inserting Bar|Beat Markers

You can insert individual Bar|Beat Markers anywhere in the timeline by setting an Edit insertion point (instead of making a selection) and then using the Identify Beat command. The ability to identify each individual beat is particularly useful when you just need to make a few corrections to tighten up the correspondence between the Pro Tools bars and beats and the audio that you are working with. If the tempo gets a little faster or slower, even within a bar, you can easily insert a new Bar|Beat Marker to make sure that Pro Tools keeps in step with your audio.

In the example shown, this drumbeat was played a little late at bar 68, beat 4, clock tick 22.

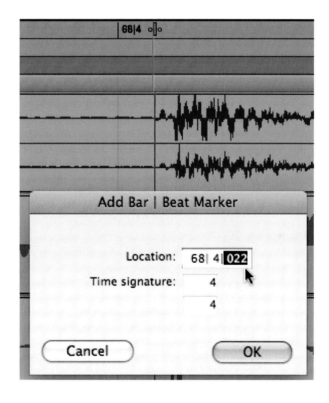

Figure 3.35
Inserting a
Bar|Beat Marker at
an Edit insertion
point.

To move the beat in Pro Tools to correspond with the actual drum beat, make an Edit insertion point using the Selector tool at the beginning of the beat and then use the Identify Beat command to open the Add Bar|Beat Marker dialog. Use this dialog to tell Pro Tools that this Edit insertion point is where this beat is supposed to be, by typing the correct bar beat and clock tick location. When you click OK in the the dialog, Pro Tools inserts a beat marker at this point along with a tempo event that adjusts the Bar|Beat location to correspond with this point.

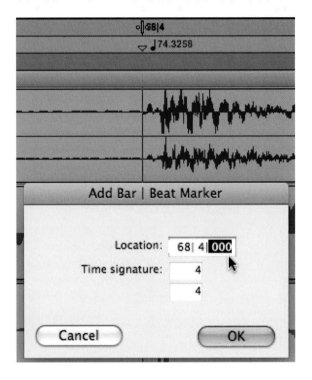

Figure 3.36
Bar|Beat Marker inserted into the Tempo Ruler to make the beat in Pro Tools correspond with the actual beat in the audio.

> **Tip**
>
> To delete a Bar|Beat Marker, just hold down the Option (Mac) or Alt (Windows) key, move the cursor over the Bar|Beat Marker, and click.

When you have your Bar|Beat Markers in place, you may need to finely adjust the locations to position these even more accurately—dragging each marker, as necessary, to align with the associated beat within the audio.

> **Note**
>
> When you drag Bar|Beat Markers back and forth along the timeline to re-position their locations, neighboring MIDI data get adjusted to align with the new tempo map.

Re-Positioning Bar|Beat Markers

Because Bar|Beat Markers are sample-based, they behave differently from tick-based Tempo events when you drag them in the Tempo ruler. When you drag a Bar|Beat Marker, its bar and beat location is dragged *with* the Bar|Beat Marker. This location does not change.

So, for example, if the Bar|Beat Marker was originally placed at 3|1|000, it will still be at this location (unless you specifically change this). However, its BPM value is recalculated along with the BPM value of the Bar|Beat Marker to its immediate left. Also, the Bar|Beat Marker's sample and SMPTE locations will be recalculated according to the new tempo of the Bar|Beat Marker. Bar|Beat Markers to the right of the dragged marker remain unchanged. Neighboring MIDI events, along with the Bars|Beats ruler, shrink or expand as necessary to adjust to the new tempo.

If you want to change the bar and beat location of a Bar|Beat Marker, you must use the Edit Bar|Beat dialog instead. If you double-click on a Bar|Beat Marker, the Edit Bar|Beat dialog opens, where you can make changes to its bar and beat location and also change the time signature if you wish.

Figure 3.37
Edit Bar|Beat
Marker dialog.

Time Operations

The Time Operations window provides four important functions controlled on separate pages. You can use this to change the meter, Insert or cut time, or move the song start position.

Setting the Meter

Pro Tools defaults to 4/4 meter, which is the time signature of most of the popular music. If you want to work in other meters such as 3/4 (waltz time) or 5/4 (like Dave Brubeck's jazz hit "Take Five") or with the changing meters that are used in various forms of music such as classical and orchestral film scores, then you will normally need to set these before you start recording.

To open the Meter Change dialog, you can either click on the "+" sign in the Meter ruler or double-click on the displayed time signature in the MIDI controls section of the Transport window.

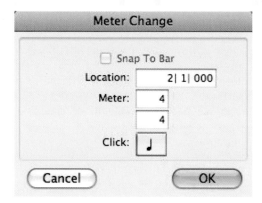

Figure 3.38
Meter Change
window.

Alternatively, you can use the Change Meter dialog, which you can select from the Event menu's Time Operations sub-menu.

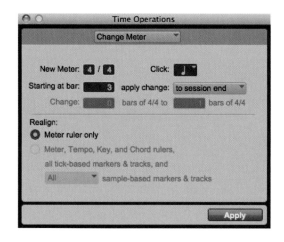

Figure 3.39
Time Operations
Change Meter
dialog.

One situation where you will need to map out tempo or meter changes is when you have recorded a musician playing without a click, perhaps to capture an improvised performance. This could involve meter changes where the musician plays a bar of 3/4 or a bar of 5/4 by just "feeling" how many beats to play in a particular section. Sometimes this is done deliberately and sometimes not, but whatever the reason, you can use the Tempo and Meter rulers to follow.

Figure 3.40
Tempo, Meter,
and Markers
rulers constructed
to follow an
improvised
performance.

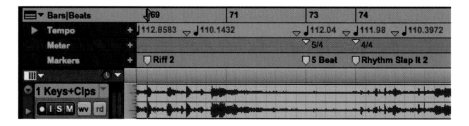

Insert or Cut Time

When you are working on your musical arrangements, it is quite likely that you will need to insert some bars or remove some bars at some time or other.

You could just move everything in your session forwards or backwards by inserting or deleting the correct number of bars, but this can be quite awkward and is not what you usually want to do. Pro Tools has features that enable you to do this conveniently and properly—adding or removing "time" to or from the timeline.

The Insert Time dialog lets you insert time from any particular start location and lets you specify the amount of time you wish to insert in terms of bars, beats, and clocks. It also moves any tempo markers forward from the insertion point.

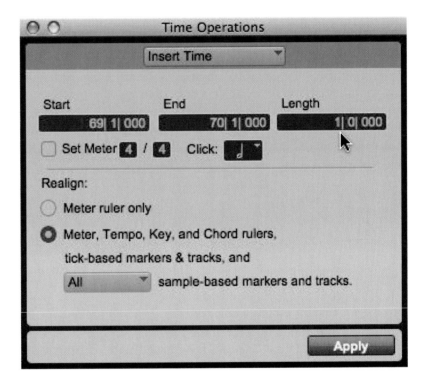

Figure 3.41 Insert Time dialog.

The Cut Time dialog lets you remove time, starting from any particular start location, and lets you specify the amount of time you wish to remove in terms of bars, beats, and clocks. This also moves any tempo markers backward from the insertion point.

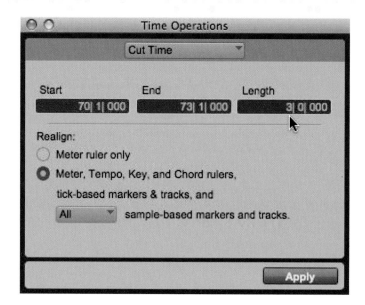

Figure 3.42
Cut Time dialog.

Move Song Start

Sometimes you may need to move the actual start point of the song. Pro Tools provides a dialog to let you specify the bar, beat, and clock location that you want to move it to. Normally, all tick-based markers and tracks will move, along with any meter and Tempo events. You can choose whether or not to move sample-based markers and track.

Figure 3.43
Move Song Start
dialog.

Key Signatures

Pro Tools provides a Key Signature ruler for adding key signatures to Pro Tools sessions. Key signatures are an important part of any musical composition that will be displayed as notation for musicians to read. Now that Pro Tools has Score Editor and separate MIDI Editor windows, it is becoming much more likely that you will need to enter key signatures as necessary into the Key Signature ruler for each Pro Tools session. The default key signature is C major, but it is highly unlikely that every piece of music you work on will be in this key!

Key signatures can be imported and exported with MIDI data. This is especially useful when exporting MIDI sequences for use in notation programs like Sibelius and can save you lots of time if you have an existing MIDI sequence to import, so as to help you get started with your music production.

All MIDI and Instrument tracks in Pro Tools now have an option called Pitched. When this is enabled, these tracks will automatically be transposed or conformed to any key signature changes—although this should be disabled on drum tracks (which should not normally be transposed). You can also make quick transpositions using the Real-Time Properties for any track or region.

The Key Signature Ruler

The Key Signature ruler lets you add, edit, and delete key signatures. You can use key signatures to indicate key and key changes in your Pro Tools session. Key signatures can also be used for various other functions such as transposing in key or constraining pitches to the specified key.

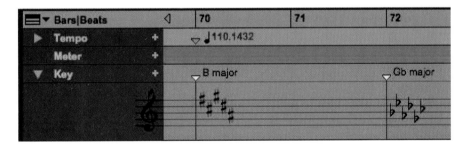

Figure 3.44
Key Signature
ruler.

To view the Key Signature ruler, select it using the Rulers sub-menu in the View menu or from the popup View selector near the top left of the Edit window. If you click on the small Show/Hide triangle at the left of the Key ruler, the Key Signature Staff will be revealed.

To the right of the ruler's name in the Edit window, there is a small "+" button that can be used to add a key signature at the current timeline position. Alternatively, you can choose "Add Key Change" from the Event menu. Using the Key Change dialog that appears, you can select the mode (major or minor), the key, its location and range, and how you want it to affect pitched tracks. You can also edit an existing key signature by double-clicking its marker in the Key Signature ruler to open the Key Change dialog.

To delete a key signature, simply Option-click (Mac) or Alt-click (Windows) the Key Signature Marker in the Key Signature ruler. Another way to do this is to make a selection in the Key Signature ruler that includes the key signature you want to delete and choose Clear from the Edit menu or press Delete.

The Key Change Dialog

Using the Key Change dialog, which opens whenever you add (or double-click to edit) a Key Signature Marker, you can choose whether the key should be major or minor, select from any of the flat keys or sharp keys, specify a range of bars throughout which the key change will apply, and choose whether or not the existing MIDI notes on pitched tracks should be transposed or constrained to the key.

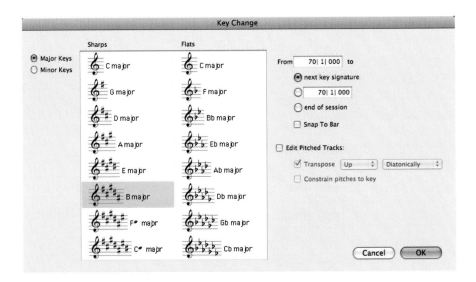

Figure 3.45
Key Change
dialog.

You can view the sharp keys in the left hand column and the flat keys in the right hand column. To the left of these columns is a pair of "radio" buttons, which lets you select major or minor keys (from the natural minor mode). To the right of the key change columns, there are various options. You can specify the Bar:Beat locations at which and until which the key change will be applied (to the next Key Signature Marker, to a selection, or to the end of the session). An option is also available to "snap" the Key Signature Marker to the bar line, which can sometimes be useful.

Enabling the "Edit Pitched Tracks" option lets you transpose existing MIDI notes on Pitched tracks up or down, either diatonically or chromatically based on the key change—or you can constrain pitches to the new key.

For example, if the session is in C major and you want to change the key to D minor, you would select "Edit Pitched Tracks," "Transpose," "Up," and "Diatonically." When you click OK in the dialog, notes on Pitched tracks will be transposed up a whole step and then the third, sixth, and seventh scale degrees will be lowered by a half step to make the key minor. Specifically, the notes C, D, and E would be transposed to D, E, and F natural. If you transpose these notes chromatically instead of diatonically, E would be transposed to F sharp instead.

So, diatonic transposition alters notes to make them conform to the type of key change that you want, whether major to minor or minor to major. Any chromatic note alterations that exist in the material being transposed, such as a flattened 7th, are left unaltered by diatonic transposition. So, for example, if you change the key from C major to E major and there is a Bb note among those you have selected for transposition, the B-flat will be transposed to D-natural—which is the flattened 7th note in the scale of E major.

The "Constrain Pitches To Key" option lets you constrain pitches to the notes of the new key. When you use this option, any diatonic pitches of the old key on Pitched tracks that are not in the new key are individually transposed to the nearest diatonic pitch of the new key, while notes that are in the new key are left untouched.

For example, when changing from C major to D major, the sequence of notes C, D, E, F, and G will become C-sharp, D, E, F-sharp, and G. The way this works is that C gets transposed to C-sharp which is in the scale of D major; D is not transposed because it is already a valid note in the scale of D major; E is not transposed because it is already a valid note in the scale of D major; F gets transposed to F-sharp which is in the scale of D major; and G is not transposed because it is already a valid note in the scale of D major.

Constraining pitches to key also constrains any chromatic pitches to the new diatonic scale. For example, going from C major to D major, the sequence of notes C, D, D-sharp, and E will be changed to C-sharp, D, D, and E. The way this works is that C gets transposed to C-sharp which is in the scale of D major; D is not transposed because it is already a valid note in the scale of D major; D-sharp is moved to D to keep it within the scale of D major; and E is not transposed because it is already a valid note in the scale of D major.

Diatonic versus Chromatic

This is most simply explained by discussing the notes on the piano keyboard: When you play the black and white keys of a piano one after the other, this is called the chromatic scale and always consists of a series of semitone intervals.

A diatonic scale is what you hear when you play the white keys of the piano one after the other. If you start on C, you get a major scale, also known as the Ionian mode. If you start on A, you get the natural minor scale (the descending form of the melodic minor scale), also known as the Aeolian mode. If you start on any other note, you get one of the other "church modes": the Dorian starting on D, the Phrygian starting on E, the Lydian starting on F, the Mixolydian starting on G, or the Locrian starting on B. You can also play a major scale starting on any black key and derive diatonic scales from this in a similar way.

Diatonic intervals are understood to be the intervals between pairs of notes taken from the same diatonic scale. Any other intervals are called chromatic intervals. Diatonic chords are understood to be those that are built using only the notes from the same diatonic scale; all other chords are considered to be chromatic. The meanings of the terms diatonic note and chromatic note vary according to the meaning of the term "diatonic scale," which is defined slightly differently by different authorities. Generally—not universally—a note is understood as diatonic in a context, if it belongs to the diatonic scale that is used in that context; otherwise it is chromatic.

Pitched Tracks

MIDI and Instrument tracks are regarded as "pitched" tracks, because MIDI note data specifies the pitch of the notes. Normally, you will want these tracks to be affected by transpositions due to key changes. But if these tracks are controlling drum machines or samplers, the opposite is true: you normally will not want these tracks to be affected by key changes because it would mess up the key mappings that cause particular drums or samples to be played.

Pro Tools allows you to override the default "pitched" behavior on individual tracks using the track's Playlist selector. Simply deselect the Pitched option (so that it has no tick mark next to it in the Playlist selector) for specific MIDI and Instrument tracks—and these won't be affected by key changes.

Figure 3.46
Using the Playlist selector to deselect the Pitched option.

Diatonic Transposition

Various diatonic transposition options are available from the Transpose dialog, which you can access using the Event Operations sub-menu in the Event menu. The Transpose Event Operations dialog allows you to transpose by octaves and semitones, or from one specified pitch to another, or to transpose all the notes you have selected to a specified pitch. It will also allow you to transpose in key, up or down, by scale steps. For example, if you have a sequence of notes that you want to double in thirds, you can simply copy this to a new track, select the MIDI notes on the new track, and transpose them up by two scale steps. In C major, for example, the sequence C, D, and E would be transposed to E, F, and G.

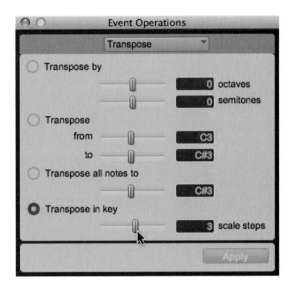

Figure 3.47
Event Operations
Transpose dialog.

You can also use the track-based MIDI Real-Time Properties or the MIDI Real-Time Properties window from the Event menu to transpose up or down by octaves and semitones, to a named pitch, or to a key.

Figure 3.48
Track-based MIDI
Real-Time
Properties
Transposition
feature.

Both of these have a Transpose Mode popup selector that lets you choose the type of transposition that you want. The default is to transpose by octave and/or semitone. You can also transpose to a specified pitch or in key.

Changes that you make using either the track-based MIDI Real-Time properties or the Real-Time Properties window are reflected in the other.

You can also apply MIDI Real-Time properties to any region on a track by selecting the region and opening the Real-Time Properties window from the Event menu.

Figure 3.49
MIDI Real-Time
Properties window
Transposition
feature.

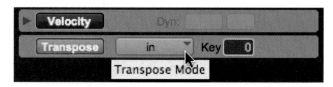

Figure 3.50
Transposing to a
specified pitch.

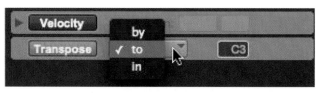

Figure 3.51
Using the
Transpose Mode
button to choose
Transpose in Key.

Figure 3.52
Region-based Real-
Time Properties
window.

Elastic Audio Region-Based Pitch Transposition

Elastic Audio can be used to change the pitch of audio regions in real time, in semitones and cents, by up to two octaves above or below the original pitch. However, just because it can change the pitch within this incredibly wide range doesn't mean that it can do this without introducing artifacts into the sound—and it does! Even when using the non-realtime X-Form rendering, you may hear warbling, phasing sounds in the pitch-shifted audio, even when you shift the pitch by as little as 1 semitone! Nevertheless, the pitch does shift, allowing you to quickly find out if things are "working" musically or not.

> **Tip**
>
> It can often be useful to shift the pitch of a musical part to assess how this would sound in a different key before going to the trouble of re-recording the part in the new key.

To apply a pitch-shift, Elastic Audio must be enabled for the track first, choosing the Polyphonic, Rhythmic, or X-Form algorithm.

Figure 3.53
Enabling Elastic Audio.

With Elastic Audio enabled on the track, select any whole region or regions within the track using the Grabber or Selector tool and then open the Elastic Properties window from the Region menu or by pressing Option-5 (Mac) or Alt-5 (Windows) on the numeric keypad. In the Elastic Properties window, you can enter the pitch shift in semitones and cents.

You can also use the Transpose window to transpose the pitch of audio regions on Elastic Audio-enabled tracks. As with the Elastic Properties window, only the regions that are completely selected will be affected. In the Transpose window, you can either transpose by octaves and semitones or you can use the "Transpose From and To" settings to transpose from one specified key note to another.

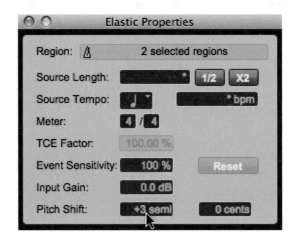

Figure 3.54
Elastic properties
window.

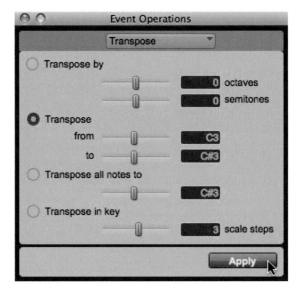

Figure 3.55
Event Operations
Transpose window.

> ## Note
>
> The "Transpose All Notes To" and the "Transpose In Key" settings can *only* be applied to MIDI notes, so these options are not active when you select audio regions.

Removing Region Pitch Shifting

If you have applied any pitch shifting to a region and you are not happy with the results, or change your mind later on, you can always remove this pitch shifting and revert the region to its original pitch.

To remove any pitch transposition applied to a selected region, you can either open the Elastic Properties window or Transpose window and remove the transposition (changing either of these, changes the other) or you can use the Remove Pitch Shift command from the Region menu—which is faster.

Note

You can only use the Remove Pitch Shift command with regions—and not with region groups. If you need to remove pitch shifting from the regions within a group, ungroup the region first, then remove the pitch shift from the individual regions, and then regroup the regions.

Elastic Audio Tempo Changes

Regions on tick-based, Elastic Audio-enabled tracks will automatically have Time Compression and Expansion (TCE) applied to alter the playback speed of the audio based on any tempo changes. So if you make simple changes to the tempo, the tempo of these audio regions will automatically follow.

Note

Tempo changes will not affect sample-based tracks, even if Elastic Audio enabled.

To Make Audio Tempo Follow Session Tempo

To have the tempo of your audio tracks speed up or slow down when you change the tempo in your Pro Tools session, simply click on the popup Elastic Audio Plug-in selector in the Track Controls area of the track containing the audio you want to affect and choose the Elastic Audio plug-in that best suits your track material and the effect you are trying to achieve.

Options include Polyphonic, Rhythmic, Monophonic, Varispeed, and X-Form. For example, with a stereo mix that you want to speed up or slow down without changing pitch, you will normally get the best results using the Polyphonic algorithm.

Real-Time Processing will be selected by default. The audio region in the track will temporarily go offline and the waveform will be grayed-out for a short time until the analysis is completed.

Make sure that the track's timebase is set to Ticks, instead of Samples. Then all you need to do is to move the Manual Tempo up or down, and the tempo will follow immediately—it couldn't get much easier than this!

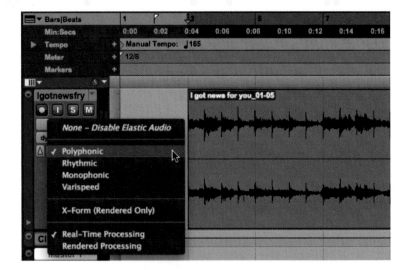

Figure 3.56
Enabling Elastic
Audio Polyphonic
Real-Time
Processing.

Figure 3.57
Changing the
track timebase to
Ticks.

Tip

To change the tempo manually, first make sure that the Conductor icon is
deselected in the Transport window. Use the mouse to point, click, and hold,
and then drag up or down on the tempo field in the Transport window.

Figure 3.58
Changing the
tempo of an Elastic
Audio-enabled
track.

> **Note**
>
> Although the tempo of the audio will change when you change the session tempo, the audio will not necessarily match the tempo of the session— unless the tempo of the audio is constant and you originally set the tempo of the session to match this.

> **Tip**
>
> To match a recording with changing tempos to the session tempo, you need to use Elastic Audio's advanced Warp matching features.

Conform to Tempo Command

A quick way to make the tempo of an audio region conform to a fixed session tempo is to use the Conform to Tempo command. Assuming that a tempo is already set in Pro Tools, all you need to do is enable one of the Elastic Audio's analysis modes on the track containing the audio region.

When you select an audio region within the track, the Conform to Tempo command becomes available in the Region menu. If you go ahead and choose the Conform to Tempo command, it analyzes the selected region to determine its tempo and duration in bars and beats. If the tempo of the region is different from the tempo of the session, Elastic Audio processing is then automatically applied to match the tempo of the region to the tempo of the session.

> **Note**
>
> The Conform to Tempo command can only be applied to regions and cannot be applied to region groups. To conform region groups to the tempo, you must first ungroup the region group, then apply Conform to Tempo to the underlying regions, and then regroup those regions.

Beat Detective

An alternative way of conforming audio to the tempo of a Pro Tools session is to use Beat Detective. Unlike Elastic Audio, Beat Detective does not use time compression/expansion. This has the advantage that it does not affect the quality of the audio that has been conformed. It may introduce audible glitches where regions have been separated, but an Edit Smoothing capability helps to alleviate this problem.

For example, Beat Detective can split beats in a drum part into separate regions that can then be moved to line up with the bars and beats grid in Pro Tools.

Take a look at the accompanying screens to see how this works with a clave pattern played using sixteenth notes. Although these are quite close to the correct bar/beat positions, they are not exact.

Using Region Separation, first capture the selection—checking that the start and end points are correct, the time signature is correct, and that you have chosen the correct beat sub-divisions (1/16 note in this case)—and then analyze this to find the beats within the selection. With this done, you can click the Separate button to split the selection into regions—each containing just one of the beats.

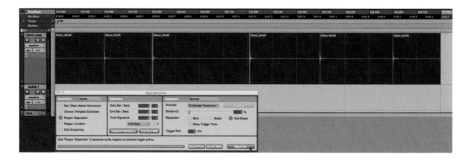

Figure 3.59
Region Separation using Beat Detective.

The next step in the process is to use the "radio" buttons at the left of the Beat Detective window to switch to Region Conform, hit the Conform button, and watch the beats/regions move to line up with the barlines!

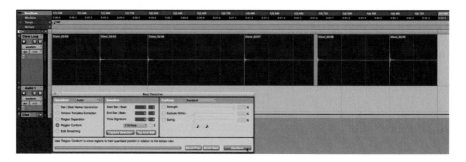

Figure 3.60
Region Conform using Beat Detective.

Typically, you will find that moving the regions will leave one or more gaps, which may be audible. Beat Detective offers a "quick fix" for this. Having separated and conformed the regions, you can switch to the Edit Smoothing operation and choose whether to simply fill the gaps (automatically trimming the region endpoints to fill the gaps) or to fill and crossfade (which also adds a few milliseconds of pre-fade before each region start point to further reduce the possibility of hearing these edit points). When you click Smooth, everything is adjusted for you automatically—which can save a lot of editing time if there are lots of gaps to fill.

Figure 3.61
Edit Smoothing
using Beat
Detective.

Beat Detective is a versatile tool, which is well worth exploring and learning how to use. Many engineers prefer to use Beat Detective's Collection mode when working with drumkits to conform the drums to the session tempo—or to have the session tempo follow the drummer's beats.

Summary

Having control of the tempo, meter, and key signature of your music is essential if you are to act as the "conductor" of the musicians, band, or orchestra that you are recording—whether real or "virtual"!

If your role is to engineer the session, then you need to be able to act on the instructions of the producer, arranger, composer, musical director, or of the musicians who are performing on the recording to establish the tempo, meter, and key. This, ideally, should be done before you start recording or as soon as possible after recording, while the people who performed (including any conductor, arranger, or producer directing the performances of the musicians) are still present at the session.

There will be times when the people involved do not want to be constrained by having to play to a click. Everyone needs to clearly understand that any further recording, editing, or mixing of the track will be much more difficult until the correct tempo(s), time signature(s), and key signature(s) have been worked out for the session. If this has to be done after the recording, it can take a significant amount of time to sort all this out. Fortunately, Pro Tools provides excellent tools to help you do this, including Identificaton of Beat, Beat Detective, and Elastic Audio.

You need to be clear about the differences between the sample-based (absolute) and the tick-based (relative) timebases and how these are displayed using the rulers. Also you need to know how and why you can change tracks to be sample-based or tick-based; the differences between Tempo events and Bar|Beat Markers; when to choose Bar|Beat or Absolute Reference Markers; and when to view the timeline using either the Linear Sample Display or the Linear Tick Display Mode. You should also be aware of how to carry out Time

Operations such as changing the meter, inserting or cutting time from your session, and moving the Song Start.

Understanding the differences between chromatic and diatonic key transposition requires some knowledge of music, but should not be beyond the capabilities of anyone who can master the other intricacies of using the Pro Tools software. Appreciating the differences and what is possible when transposing MIDI and audio regions also requires understanding of music and of MIDI itself. There are many good books published elsewhere that cover these topics more thoroughly than is possible here.

Elastic Audio and Beat Detective are great "tools" to have at your disposal when you need to make your audio follow the session tempo, or to edit or move notes to align with others and are well worth spending time with so that you are familiar with all their applications.

Anyway, if you are still unsure about any of these things, please go back and review this chapter, and make sure that you try everything for yourself on your own system. For some people, myself included, it can take a little while before the importance of all these things becomes apparent. When this does happen—learning follows much more quickly!

In This Chapter

Working with MIDI

Introduction

There are many Pro Tools users who will use MIDI in their music production work—even when starting on a project that may end up as audio-only by the time that it is finished. It's just great to be able to try ideas out using piano, bass, drums, percussion, orchestral sounds, or vocal samples when you are working on a new composition or arrangement. Of course, there is plenty of music being recorded today using MIDI instruments, virtual instruments, synthesized and sampled sounds, very heavily or even exclusively.

Controlling a MIDI Rack from Pro Tools

Before we start using virtual instruments, let's take a quick look at how things work with a rack of real MIDI hardware. Before virtual instruments were available, MIDI musicians used real instruments—keyboard synthesizers such as the Yamaha DX7 and 19 in. rack-mountable MIDI "modules" such as the Roland MKS80. There are still lots of these instruments "out there," so it's quite likely that you will come across these at some time or other.

For example, I still have a Yamaha DX7II synthesizer, a KX88 weighted MIDI Master Keyboard, and a flightcase containing a Yamaha TX816, TX802, and TX81Z, a Roland MKS80 and MKS70, with a MOTU MIDI Express XT interface to connect these to my computer and a Yamaha 03D mixing console that I use as a submixer. These synthesizers are my 1980s legacy, as anyone who is familiar with these instruments will instantly realize. Of course, Yamaha and Roland are still making hardware synthesizers today, along with more recent arrivals such as Novation and Clavia.

Computer MIDI Setup

Before you can use any MIDI hardware with Pro Tools, you will need to set up the MIDI connections to and from your computer first. On the Mac, for example, you can open the MIDI Setup application using the "MIDI Studio…" menu option available from the Setup menu in Pro Tools. Here, you can add a device corresponding to each actual device in your MIDI rack and hook these up to your MIDI interface to reflect the actual cable connections you have made.

> ## Tip
>
> To make sure everything is working OK, just click on "Test Setup" in the Apple MIDI Setup MIDI Devices window, then click on each device in your setup. This sends a bunch of MIDI notes to the device under test. You should see MIDI input indicators flash on each device that you test and you should hear audio if the device's audio outputs are connected to a monitoring system. Alternatively, you can plug a set of headphones into each device that you are testing to make sure it is playing OK.

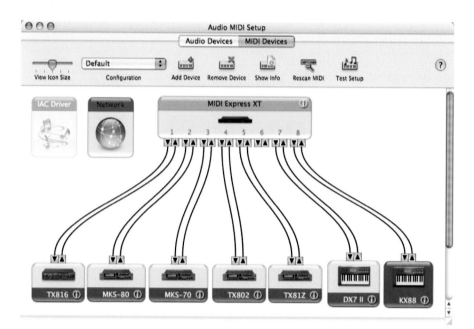

Figure 4.1 MIDI Devices "page" in the Apple Audio MIDI Setup utility.

On the PC, when you choose MIDI Studio Setup from the Setup menu, you can create an entry for each "Instrument" that you have in your setup in a similar way. Pop-up selectors let you choose from a list of popular MIDI devices and models, and choose the Input and Output ports. You can also choose which MIDI channels to send and receive on for each device.

Pro Tools MIDI Setup

When you are sure that everything in your MIDI rack is working OK, then it's time to start working with Pro Tools. A good way to work with a rack of MIDI gear like this is to set up a Pro Tools session with a MIDI track for each instrument, an Aux track to monitor each instrument, and an Audio track to record each instrument into when you have finished recording and editing the MIDI track.

You will need to route the outputs from each hardware MIDI device into available inputs on your Pro Tools audio interface and choose these as the inputs to the Auxiliary Input tracks that you have created to monitor (i.e., listen to) each MIDI device. If you don't have enough inputs, you will either have to swap cables and record instruments individually or use a submixer.

When you are ready to record a MIDI track as audio, simply route the output of the Aux track that you are using to monitor this into the Audio track that you have created to record it. When you finish recording the MIDI track as audio, you can make the Auxiliary track inactive, mute the MIDI track, hide both of these tracks, and just play back the audio track. Of course, if you need to make any changes at any point in the future, you still have the original MIDI and Aux tracks to return to.

> **Note**
>
> You can use Instrument tracks instead of MIDI and Aux tracks, but let's do things this way first—so that you get to see exactly how everything works.

Each MIDI device that you are using should have its own MIDI track in Pro Tools with the track output set to control that device. Typically, you would use a MIDI Master Keyboard (I'm using a Yamaha KX88) as the main input device to record into each track.

> **Note**
>
> Keep in mind that many MIDI devices can operate as "multitimbral" instruments. This means that they can be set up to operate as though they contain two or more separate MIDI devices—each of which is controlled by a different MIDI channel. You may encounter devices that have as many as 16 multitimbral instruments, or even 32, in which case they will have two separate 16-channel MIDI interface ports. They may also have multiple audio outputs. The Yamaha TX816 has eight totally separate outputs, for example, so I use an 8:2 audio submixer with this.

Figure 4.2 A Pro Tools Session showing a MIDI track with an Auxiliary Input to monitor the audio from a synthesizer and an Audio track into which this audio can be recorded. To record the MIDI as audio, the outputs of the Auxiliary Input need to be routed to the inputs of the associated Audio track.

In my rig, the Yamaha DX7II and the Roland MKS80 can each respond to two MIDI channels, while the others can respond to up to eight MIDI channels. I say "up to" eight MIDI channels because it is possible to set these devices to play all their multitimbral instruments from the same MIDI channel or from a smaller combination of MIDI channels. For example, the Yamaha TX816 contains eight DX7 synthesizer modules in one rack: each module can be set to whatever MIDI channel you like, so you can use them in pairs with each pair set to a different MIDI channel, giving you four different sets of sounds, or you can set all eight modules to the same MIDI channel and use all eight to layer up one very big, complex sound.

> **Tip**
>
> It is also possible to achieve the same result by leaving the MIDI modules in your rack on separate MIDI channels, then routing the outputs from your MIDI tracks in Pro Tools to any number of these modules. This way you can layer up sounds any way you like using your various instruments.

MIDI Thru Preferences

I have two keyboards in my rig—the KX88 and the DX7II. The first time that I set everything up, I could not work out why whenever I played the DX7II keyboard it would also play the TX816. After all, I had carefully set the MIDI track inputs to use either the KX88 or the DX7II. So what was going on here?

It took me a few minutes of head scratching before I figured out that there must be a default preference that was causing this. Sure enough, in the MIDI Preferences, I found an option to set a Default Thru Instrument. This was set to "Follows First Selected MIDI Track," and the first MIDI track in my Pro Tools session, the TX816, was selected (highlighted).

So that's why the TX816 played back even if I was playing on the DX7II keyboard! With this preference set to "None," the problem immediately disappeared.

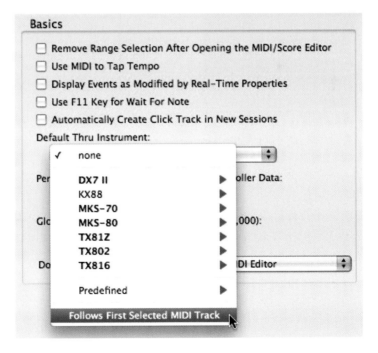

Figure 4.3 MIDI Preferences "page" showing the Default MIDI Thru Instrument settings.

Initial MIDI System Setup Data

It's always a good idea to put initial patch changes for your MIDI gear and any other setup data (such as controllers like pitch bend and modulation set to sensible defaults) into a couple of setup bars that play back before any music plays. This way you can make sure that the MIDI devices in your rig are all ready to play back by the time the first bar of music is reached. Resetting the pitch bend and the modulation to their default positions would be two of the most basic things that you might do here—in case these controls have been left in some other state before you start to play back your new sequences.

You can even embed the actual patches and other data for your MIDI gear into each MIDI track—storing this as MIDI System Exclusive (SysEx) data. Be aware that sending out SysEx data will disrupt other MIDI data such as notes and timing information, so you may need to wait until several bars have passed before you start the rest of your MIDI playback. To avoid this disruption, you could put the SysEx data in dedicated tracks, mute all the other tracks temporarily, play back the tracks containing the SysEx data to set up your MIDI rig, then mute all the Sysex tracks and unmute the tracks you want to play back.

To record SysEx into Pro Tools, enable a MIDI track to record, make sure that the SysEx data is not being filtered out by the Pro Tools MIDI Input Filter, click "Wait for Note" in the Transport window, make sure that you are at the beginning of your track by pressing the "Return" key, then initiate the SysEx transfer from your MIDI device. When this has finished, hit the Spacebar or click "Stop" in the Transport window to stop recording. The SysEx data will appear as a MIDI region in the track's playlist and in the MIDI regions list. SysEx messages are also displayed in the MIDI Event List where they can be copied, deleted, or moved as necessary.

Figure 4.4 MIDI Event List window showing a SysEx Event.

Note

Although Pro Tools will record and playback SysEx data, it will not let you edit this data or write in data manually—unlike Digital Performer, for example, which does offer these more advanced features.

Saving Templates

Having set up and tested every device in your rig, with MIDI, Auxiliary Input, and Audio tracks for these created in Pro Tools, you should save your session as a template file and keep it somewhere safe—either in the system or in a project folder. If you save it with the system, it will always be available as a template to choose from when you create a new session.

Import Session

If you have already opened a Pro Tools session and you realize that there are some tracks set up in other sessions, or in a Template session, you can use the Import Session Data command to bring in tracks that you want to use from the other session into your current session.

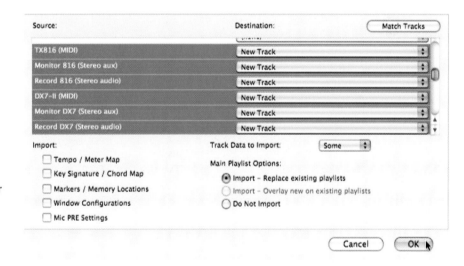

Figure 4.5 Lower part of Import Session Data dialog showing tracks being imported.

Recording Audio from MIDI Hardware Devices

When you have finished recording and editing your MIDI tracks, it is always best to record these as audio. There are several reasons for this, but the fact that you can disconnect the MIDI hardware and still hear the audio playback is probably the most important reason that everyone will be able to appreciate. Another major advantage is that you can play back the session on any other Pro Tools system, such as from a laptop when you are away from your studio, without having to carry the MIDI hardware with you.

While you are working on the MIDI sequences, you need to monitor the audio coming from the MIDI device (synthesizer, sampler, drum machine, or whatever). The best choices here are to use either an Auxiliary Input or an Instrument track, neither of which will use up any of the available "voices" that you may need for your audio tracks. When you need to record the audio from the MIDI device, you will need an Audio track with its input set to receive the output from the Auxiliary Input or Instrument track that you are using to monitor the audio from your MIDI device. Take a look at the accompanying screenshot to see how this looks in the Pro Tools Mixer.

Figure 4.6 Recording audio from a Yamaha TX816 Synthesizer.

> **Note**
>
> You can also record audio from an external MIDI device by setting the Audio track's input to the same audio input as the Auxiliary Input (or Instrument track) being used to monitor the external MIDI device. This second method avoids any additional latency associated with bussing, but don't forget to mute the Auxiliary Input (or Instrument track) used for monitoring while recording from the same audio path to the Audio track. Alternatively, mute the Audio track. Otherwise, you will get problems due to double monitoring of the same signal via the two separate signal paths through the Mixer.

> **Tip**
>
> When you have finished recording the audio from a MIDI device, you can mute the MIDI track, make the Auxiliary track inactive, and hide both of these. You can always reverse this procedure if you need to make any changes later on in your production workflow.

Compensating for Triggering Delays

All MIDI devices are not equal. When a particular synthesizer, such as a Yamaha DX7, receives MIDI data, it will take a few milliseconds before it plays the sound. Send the same data to a Roland synthesizer, and this will take a different number of milliseconds to respond. Try the same thing with an Akai sampler and an EMU sampler. Almost certainly, the response times before they play the audio will be different.

If the response time is very fast, say 5 ms or less, and depending on the type of sound the synthesizer is playing, this may not be a problem. String sounds usually have a slow attack time, for example, so you would probably not hear anything "wrong." Where it does become a problem is when you are working with a MIDI device that has a very slow response time, or when you are working with sounds that have very fast attack times, such as percussion instruments. And if you have some audio in Pro Tools that should play exactly at the same time as the audio coming from an external MIDI device, this can exacerbate the problem. Remember that it usually takes at least 5 ms to trigger the notes from an external MIDI device, and it could take even longer, depending on the device. So if you have a kick drum that is being played by an audio track in Pro Tools and a kick drum that is being played by a MIDI device triggered by MIDI notes coming from Pro Tools, you would hear a "flam" (a delay) between the two kick drums.

To compensate for delays caused by the time it takes to trigger events on a MIDI sampler or synthesizer, you can offset individual MIDI and Instrument track offsets in Pro Tools.

> ## Note
> You can measure the latency (delay) for a MIDI device assigned to a MIDI track by recording its audio output back into Pro Tools. Then, compare the sample locations for the recorded audio events against the original MIDI notes to calculate the latency (delay) and use this value when setting the MIDI Track Offset.

To configure a MIDI or Instrument track offset for a track, choose MIDI Track Offsets from the Event menu.

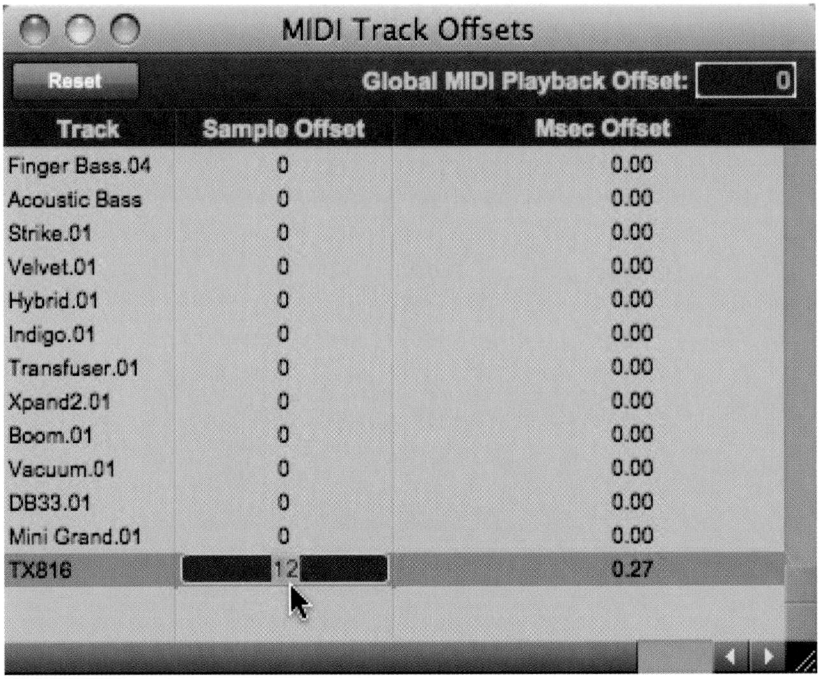

Figure 4.7
Entering individual MIDI/Instrument Track Offsets.

When the window opens, click in the Sample Offset column for the MIDI or Instrument track and enter a number between –10,000 and +10,000 to specify the offset in samples. Negative values cause the MIDI or Instrument tracks to play back earlier than the audio tracks; positive values cause the MIDI or Instrument tracks to play back later. The Msec Offset column shows the equivalent offset in milliseconds. This value updates when a new value is entered in the Sample Offset column, although it cannot be edited. Just press Return (Mac) or Enter (Windows) on your computer keyboard to accept the entered offset value. A click on the Reset button in the upper left of the window resets the offsets for all the MIDI and Instrument tracks.

Tip

You can also apply a delay in milliseconds or in ticks using the MIDI Real-Time Properties window to offset MIDI or Instrument tracks. Another possibility is to use this feature to advance (as opposed to delay) a track containing slowly evolving sounds, such as strings, so that the full impact occurs sooner.

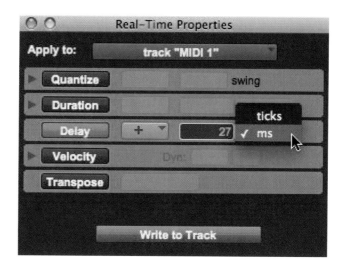

Figure 4.8 The MIDI Real-Time Properties can be used to delay a track.

MIDI Beat Clock

You may have MIDI devices connected to your Pro Tools system that can use MIDI Beat Clock to synchronize with the tempo of the Pro Tools sequence. This is often the case with drum machines or with synthesizers that have arpeggiators. Delay units are another example where the timing of the delay repeats can be synchronized to the session tempo in Pro Tools.

Note

Many plug-in instruments and effects can also be synchronized using MIDI Beat Clock. Most of these automatically configure to receive MIDI Beat Clock from Pro Tools. Plug-ins that do not self-configure are detected when inserted, then listed in the MIDI Beat Clock dialog where you can select the ones that you wish to synchronize manually.

MIDI Beat Clock Offsets

As previously mentioned, MIDI hardware devices typically take a number of milliseconds to respond after receiving MIDI note and performance data. The same thing happens with MIDI Beat Clocks. Pro Tools lets you set a MIDI Beat Clock offset for each port or external device on your MIDI interface (or for any plug-ins that do not self-configure). If you know how many milliseconds a particular device takes to respond, you can compensate for this latency (delay) by entering a negative offset value in samples for the device (or the port that it is connected to).

You can access the MIDI Beat Clock dialog from the MIDI submenu in the Setup Menu. Here, you can enable MIDI Beat Clock transmission for any of the devices listed. You can also enter negative offset values (such as –141 samples) for each device or port that you enable.

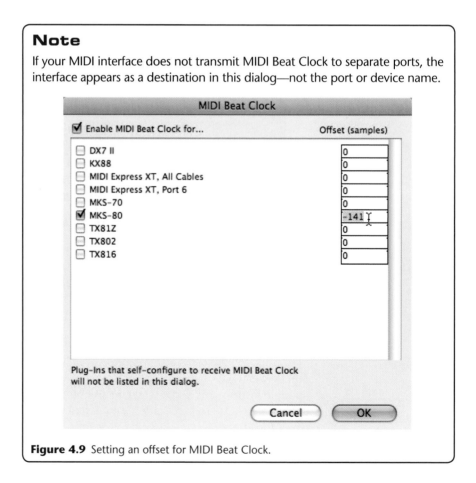

> **Note**
>
> If your MIDI interface does not transmit MIDI Beat Clock to separate ports, the interface appears as a destination in this dialog—not the port or device name.

Figure 4.9 Setting an offset for MIDI Beat Clock.

Measuring Beat Clock Latency

To discover the correct MIDI Beat Clock offset for an external MIDI device, you can record the audio output from a synchronized external device, such as a drum machine, and see any delay by examining the waveform in the Pro Tools Edit window.

An efficient way to do this is as follows: first enable both Snap to Grid and Show Grid. Then, using the Selector tool, place the edit cursor on a beat Grid line immediately before a prominent transient in the waveform that should fall on this beat (but is a little late due to the latency of the MIDI device). Enable "Tab To Transients" then press the Shift and Tab keys to make a selection from the first edit point (on the Grid line) to the transient. The duration of this selection will be the approximate amount of latency for your external MIDI device. If you then switch the Main Time Scale to Samples, the length of the selected region displayed in samples is the MIDI Beat Clock latency for that device.

> **Note**
>
> Because MIDI is not sample accurate, you may want to make several measurements at different grid locations and average them to come up with the best value for the MIDI Beat Clock Offset.

Having ascertained this information, open the MIDI Beat Clock dialog and enter the Sample Offset value as a negative number so that Pro Tools will compensate for this latency when you make future recordings using this device in this session.

> **Tip**
>
> You should carry out this procedure once for all the MIDI Beat Clock-enabled devices in your setup, and keep a record of these values so that you can re-enter the correct sample offsets for future sessions.

How Delay Compensation Works with MIDI

Pro Tools provides automatic Delay Compensation for DSP delays that can occur when you are using plug-in or hardware inserts or as a result of routing audio signals around the mixer using busses or sends. You should normally have Delay Compensation enabled to maintain phase coherent time alignment between tracks that have plug-ins with differing DSP delays, tracks with different mixing paths, tracks that are split off and recombined within the mixer, and tracks with hardware inserts.

It is important to be aware that when Delay Compensation is active, a MIDI event that is recorded to sound "in time" with delay-compensated material is actually recorded late by the length of total delay in effect. To compensate for this, MIDI events are shifted back in time by the total session delay following each MIDI recording pass. However, during some recording scenarios, you may need to override the Delay Compensation for particular tracks.

You should also be aware that Delay Compensation is suspended while recording virtual MIDI Instruments. To allow latency-free monitoring of virtual instrument plug-ins during recording, Pro Tools automatically suspends Delay Compensation through the main outputs of the Instrument track, Auxiliary Input, or Audio track on which an instrument plug-in is inserted when a MIDI or Instrument track that is routing MIDI data to this instrument plug-in is record-enabled. Obviously, this is necessary: otherwise, the musician playing the MIDI keyboard or other controller would hear a delayed version of the audio from the Instrument plug-in, which would make it difficult to play in time. Of course, during playback of the virtual MIDI instrument, when all the MIDI and audio connections take place inside Pro Tools, Delay Compensation is restored.

Because this automatic behavior is not always appropriate, manual intervention may be necessary, for example, when using MIDI and Audio processing plug-ins. Some instrument plug-ins, or audio processing plug-ins such as Avid's Bruno and Reso, let you process audio, while allowing MIDI data to control processing parameters or while using MIDI Beat Clock to synchronize with the session tempo. By default, when you record-enable a MIDI or Instrument track that is controlling an audio processing plug-in, the track the plug-in is inserted on will go into Low Latency mode—with Delay Compensation suspended—so the processed audio will play early. However, the audio output from the processing plug-in should be delay compensated just like the other tracks in the session if it is to be correctly time-aligned, while you are listening to this. To achieve this, you will need to override the suspension of the Delay Compensation as explained below:

If you are recording while using a MIDI controlled plug-in on an audio track, you can Command-Control-click (Mac) or Start-Control-click (Windows) on the Track Compensation indicator to override the suspended Delay Compensation so that the audio track will correctly apply Delay Compensation. The Track Compensation indicator will turn blue to warn that the Delay Compensation has been forcibly enabled on the track.

If you are recording while using a MIDI controlled plug-in on an Auxiliary Input, you will need to bypass the automatic delay compensation and manually apply the total system delay. Command-Control-click (Mac) or Start-Control-click (Windows) the Track Compensation indicator so that the Auxiliary Input will bypass the Delay Compensation. The Track Compensation indicator will turn gray to indicate that Delay Compensation for the track is bypassed and that no delay is being applied to the track. Then you will need to check the Session Setup window to discover the total system delay and manually enter this total system delay into the User Offset field.

Using Delay Compensation View

There are three numeric display fields in the Delay Compensation view: the Delay indicator, the User Offset field, and the Track Compensation indicator. You can bypass any of these by Command-Control-clicking (Mac) or Start-Control-clicking (Windows) on the display field, in which case the display field will turn gray.

Normally, the alphanumeric characters in these fields are colored green, but when the amount of plug-in and hardware insert delay on the track exceeds the available amount of Delay Compensation, the color turns red. Also, the color turns orange on the track reporting the longest plug-in and hardware insert delay in the session.

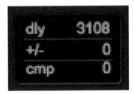

Figure 4.10
Delay
Compensation
view.

Delay values are shown in samples or milliseconds, depending on the Delay Compensation Time Mode setting in the Operation Preferences window.

Figure 4.11
Delay
Compensation
Time Mode
preference.

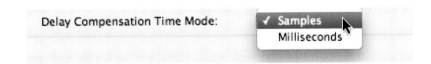

Delay Indicator

The Delay indicator reports the total delay due to inserting TDM and/or RTAS plug-ins on the track, plus any delays due to bussing audio to and from any external hardware. When Delay Compensation is activated, the delay compensation engine applies a compensating amount of delay for the track corresponding to the figure reported here.

What to Do If Track Delay Exceeds the Automatic Delay Compensation Limit!

If the total delay on a track exceeds the Delay Compensation limit, the Delay Compensation View will turn red to warn you of this. In this case, the automatic Delay Compensation can be bypassed and the track can be manually time-aligned instead.

1. Bypass the reported delay for the track by Command-Control-clicking (Mac) or Start-Control-clicking (Windows) on the Delay indicator, which turns gray.
2. Note the exact amount of delay reported in the track's Delay indicator.
3. Manually nudge any audio data on the track earlier by that amount.

User Offset

The User Offset field lets you adjust track delays manually while Delay Compensation is enabled. You can use this to adjust the way a track's timing feels by adding or subtracting from the amount of delay applied by the automatic Delay Compensation.

> **Note**
>
> A positive number (with or without the "+" modifier) sets a positive delay, making the track sound later in time, whereas a negative number (with the "–" modifier) sets a negative delay, making the track sound earlier in time.

Track Compensation

The Track Compensation indicator shows the amount of Delay Compensation that is being applied to the track.

When an Instrument track is record-enabled, its Delay Compensation is automatically suspended to provide low-latency monitoring, so the Track Compensation indicator shows zero delay.

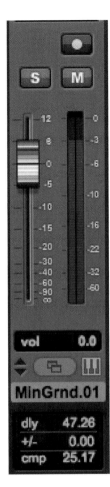

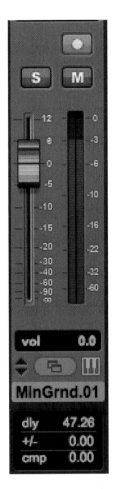

Figure 4.12 Instrument track showing Delay Compensation view—the Track is *not* record-enabled so Delay Compensation is applied.

Figure 4.13 Instrument track showing Delay Compensation view—the Track *is* record-enabled, so Delay Compensation is suspended.

Instrument Tracks

Instrument tracks make your session a little more convenient to set up if you are planning to record just one MIDI part for each MIDI or virtual instrument. An Instrument track is a substitute for a MIDI track paired with an Auxiliary Input track used to monitor the audio output from a MIDI hardware instrument or a virtual software instrument. So, an Instrument track can do everything a MIDI track needs to do and everything an Auxiliary track needs to do when recording and playing back MIDI data and monitoring the audio data from a MIDI or virtual instrument.

Like a MIDI track, the Instrument track lets you record MIDI data onto the track. Like an Auxiliary Input track, the Instrument track lets you monitor audio coming from an external MIDI device or lets you monitor audio from an inserted virtual instrument plug-in.

If you are recording MIDI drum parts, you may want to record each drum onto a separate MIDI track so that you can edit these separately. So you might have tracks for bass drum, snare drum, hi-hats, toms, and cymbals. If this is the case, there is not much advantage to using an Instrument track: you might as well just use an Auxiliary Input track to monitor the audio from the drum-machine, sampler, or virtual instrument. Of course, you could use an Instrument track for one of these instruments, such as the bass drum, then use separate MIDI tracks for the snare and other drums, with the MIDI output of each track set to control the appropriate drum sound from the MIDI or virtual instrument that is being monitored via the Instrument track.

Recording MIDI onto a MIDI or Instrument Track

To record MIDI data, first you need to create a MIDI or Instrument track or choose an existing one. Then, assign the track input and output to accept input from the MIDI keyboard or controller that you will be using and route the track's MIDI output to the MIDI device that you wish to control.

In the Mix window, you can click the track's MIDI Input Device/Channel Selector and assign a device and channel from the pop-up menu. If you have more than one keyboard or other MIDI device hooked up that you want to play and create MIDI data with, you can use the "All" option to allow input from all your connected MIDI devices. I have two MIDI keyboards, for example, one with a weighted action for playing piano sounds and one with a non-weighted action for playing synthesizer sounds.

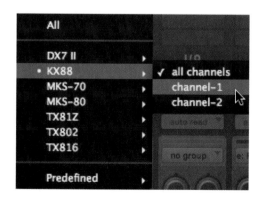

Figure 4.14 The MIDI Input pop-up selector showing the available sources in my test session.

You should also assign the MIDI output of the track to the MIDI instrument—the actual hardware or the "virtual" software—that you will be using. You may also need to enable the MIDI Thru item in the Options menu.

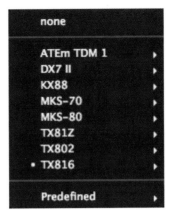

Figure 4.15 The MIDI Output pop-up selector showing the available destinations in my test session.

> ## Tip
> Remember that when you are recording MIDI onto Instrument tracks, the MIDI Input and Output pop-up selectors are located at the top of the Instruments section rather than in the track I/O section. This is because Instrument tracks can be used to accept audio input from external MIDI devices as well as from virtual instrument plug-ins. So, the track I/O section is used to set up the input and output paths for this audio—not to set up MIDI I/O.

Record Mode

To enter Record mode, first you need to Record-enable the track and make sure that it is receiving MIDI from your keyboard or other controller. The track meters will indicate MIDI input when you play your MIDI device if everything is set up and working OK.

In the Transport window, click Return To Zero to start recording from the beginning of the session—or simply press the Return key on the computer's keyboard. If you prefer, you can start recording from wherever the cursor is located in the Edit window, or, if you select a range of time in the Edit window, recording will start at the beginning of this selection and will automatically finish at the end of the selection.

Click Record in the Transport window to enable Record mode then click Play in the Transport window or press the Spacebar to actually begin recording.

> ### Tip
>
> You can start recording by pressing and holding the Command key (Control key on the PC) then pressing the Spacebar on your computer keyboard—which is even faster than using Steps 3 and 4 above. And if you prefer to hit a single key, just press function key F12 or press 3 on the numeric keypad (when the Numeric Keypad mode is set to Transport) instead.

Figure 4.16 Transport Window with Record and Play buttons engaged.

> ### Note
>
> If you want the first MIDI note that you play to start the recording, you can enable the Wait for Note button in the Transport window (this is the first of the four buttons in the MIDI controls section and has a MIDI socket icon). If you want to hear a metronome click while you record, make sure that the click is set up correctly and enable the Metronome button. To hear a countoff, enable the Countoff button. These buttons will each turn blue when enabled.

When you have finished recording, click Stop in the Transport window or press the Spacebar. The newly recorded MIDI data will appear as a MIDI region on the track in the Edit window and in the Regions List. To see the notes, choose the Notes view from the pop-up selector in the Track controls area at the left of the Edit window.

Figure 4.17 MIDI Data in Notes view in the Edit Window.

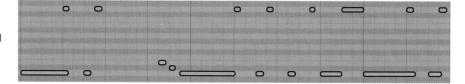

Canceling or Undoing Your MIDI Recording

If you mess up while you are recording, which can easily happen before you get used to playing the part you want to record, you can stop the recording and simultaneously discard any stuff that you have recorded (including multiple takes when in Loop Record) using one simple keyboard command. On the Mac, just press the Command key (Mac) or Control key (Windows) together with the period (.) key, which is known as the "full stop" key in UK English.

If you have stopped recording and then decide that you want to discard the last take, you can use the standard Undo (last action) command by selecting this from the Edit menu or pressing Command-z (Mac) or Control-z (Windows).

Loop Recording MIDI

There are two ways to loop record with MIDI—either using the normal non-destructive Record mode with Loop Playback and MIDI Merge enabled for drum-machine style recording or using the special Loop Record mode to record multiple takes on each record pass—as when loop recording audio.

To set up drum-machine style loop recording, where each time around the loop you record extra beats until you have constructed the pattern you want, you need to enable the MIDI Merge function by clicking its button in the Transport window or in the Edit window's MIDI Controls section.

Figure 4.18 The MIDI Merge button.

Make sure that "QuickPunch," "TrackPunch," "DestructivePunch," "Loop Record," and "Destructive Record" are not selected in the Operations menu, but do select "Loop Playback" so that the loop symbol appears around the "Play" button in the Transport window. You should also disable "Wait for Note" and "Countoff" in the Transport window.

Select "Link Edit and Timeline Selection" from the Options menu, then make a selection in the Edit window to encompass the range that you want to loop around. If you want to hear the audio that plays immediately before the loop range as a cue, you will need to set a preroll time. So, for example, if you set a 2-bar preroll, you would hear your session play back from 2 bars before the loop range, then it would play around the loop until you hit "Stop." Each time through the loop, you can add more notes until it sounds the way you want it to, without erasing any of the notes from previous passes through the loop—just like using a drum-machine.

> ## Tip
>
> To switch to a new record track, press Command (Mac), or Control (Windows), and press the Up/Down Arrow keys to record-enable the previous or next MIDI or Instrument track.

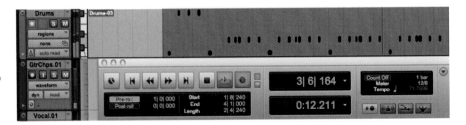

Figure 4.19 Loop recording—drum-machine style.

If you record MIDI using the Loop Record mode instead, new regions are created each time you record new notes during successive passes through the loop. This time, select "Loop Record" and "Link Edit and Timeline Selection" from the Options menu and deselect "Loop Playback." With "Loop Record" enabled, a loop symbol appears around the "Record" button in the Transport window. Make a selection in the Edit window to encompass the range that you want to loop around and set a preroll time if you need this. Start recording and play your MIDI keyboard or other MIDI controller. Each time around the loop, a new MIDI region is recorded and placed into the Edit window, replacing the previous region. When you stop recording, the most recently recorded of these "takes" is left in the track and all the takes appear as consecutively numbered regions in the MIDI Regions List.

The easiest way to audition the various takes is to make sure that the take currently residing in the track is selected, then Command-click (Control-click in Windows) on the selected region with the Selector tool enabled. A pop-up menu, the "Takes List," appears containing all your recorded takes. You can choose any of these to replace the take that currently appears in the track.

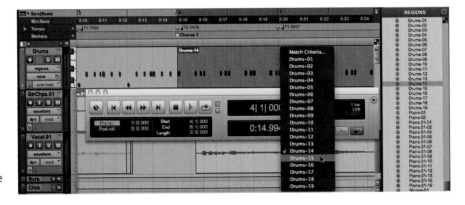

Figure 4.20 Auditioning "takes" in Loop Record mode. These are also accessible from the Regions List.

Recording Audio from a Virtual Instrument

You should always record your virtual instruments as audio as soon as possible in the production process. There are two reasons for this. First, you can remove the virtual instrument so that it is no longer using DSP resources. Second, and possibly even more importantly, if you take your Pro Tools session to another studio that doesn't have the particular virtual instrument that you used, you would not be able to recreate the original sound that you had—but if you have recorded the audio, then you have it!

You may think that the way to record a virtual instrument is to simply use an Audio track instead of an Auxiliary track to monitor your virtual instrument so that you can just put this Audio track into record while playing back the MIDI into the virtual instrument. Unfortunately, Pro Tools does not make it as simple as this. If you try this scenario, you will soon discover that the audio from the virtual instrument cannot be recorded onto the same Audio track that is monitoring the virtual instrument. The solution is to route the output from the Instrument (or Auxiliary) track used to monitor the virtual instrument to the input of an Audio track using an internal bus; then, use this Audio track to record the audio from the Instrument track.

Figure 4.21
Recording audio from a Virtual Instrument into an Audio Track: output from Instrument track routed to input of Audio track.

> **Tip**
>
> After you have recorded the audio from your virtual instrument, you can make the Instrument track inactive to release the DSP resources used and hide the track: you can always reveal the Instrument track and make it active again if you want to make some changes later on.

Editing MIDI

Pro Tools lets you edit MIDI data alphanumerically or graphically, according to your preference. You can use the standard Cut, Copy, Paste, and Clear commands to manipulate MIDI data, and Pro Tools also has "Special" versions of these commands that provide extra functions for MIDI data. There are many useful keyboard commands and display zooming features that it will pay you to become familiar with, so you should practice using these until they become "second nature." The MIDI Real-Time Properties feature lets you apply changes to regions or tracks with a minimum of fuss and the MIDI Operations windows let you transform MIDI data in a variety of ways, so time spent learning these features will be well rewarded.

Bars, Beats, and Clock "Ticks"

When you are working on music, you will normally want to display the Bars:Beats ruler so that you can edit according to the bar lines and correct note positions. MIDI data is recorded into Pro Tools with a very high degree of accuracy. The internal "clock" to which MIDI is resolved has an incredible 960,000 pulses per quarter note (PPQN) resolution. However, when the Time Scale is set to Bars:Beats, the *display* resolution in Pro Tools is 960 PPQN, which provides more manageable numbers to work with.

When you are editing MIDI data using bars and beats, there are several sets of circumstances where you may want to specify tick values. For example, when placing and spotting regions, when setting lengths for regions or MIDI notes, when locating and setting play and record ranges (including preroll and postroll), when specifying settings in the Quantize and Change Duration windows, and when setting the Grid and Nudge values.

Bars are subdivided into beats, and beats are subdivided into "ticks," so the locations of MIDI notes are specified in terms of bars, beats, and ticks. Take a look at the accompanying screenshot to see a selected MIDI note in the Edit window with its location and length also displayed in the MIDI Event List. The selected note starts at bar 11, beat 1, tick 000 and the note lasts for 1 bar, so it ends at bar 12, beat 1, tick 000.

Figure 4.22
A MIDI note displayed graphically in the Edit window with its parameters displayed alphanumerically in the MIDI Event List to the right.

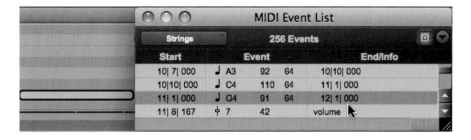

It helps to become familiar with the numbers of ticks that correspond to the main normal, dotted, and triplet note lengths. For a half note, these are 1920, 2880, and 1280 ticks. For a quarter note, these are 960, 1440, and 640 ticks. For an eighth note, these are 480, 720, and 320 ticks. For a sixteenth note, these are 240, 360, and 160 ticks. For a thirty-secondth note, these are 120, 180, and 80 ticks, and for a sixty-fourth note, these are 60, 90, and 40 ticks.

Shortcuts to Adjust Grid and Nudge Values

When you are working on your MIDI sequences, there are many times when you will want to change the grid values, so you can snap regions or notes to bars or to beats or whatever, or to change nudge values so you can nudge by smaller or larger increments.

If you are using a Mac, simply hold down the Control and Option keys and use the "+" and "−" keys on the numeric keypad to adjust the grid values. If you are using Windows, hold down the Start and Alt keys instead.

Obviously it is helpful to be able to see the grid lines in the Edit window, so make sure that you have enabled the "Draw Grid Lines in Edit Window" preference in the Display section of the Preferences window.

> ### Tip
> A useful shortcut to toggle the grid lines on and off is to Command-click (Mac) or Control-click (Windows) on the name of the Main ruler, near the top left of the Edit window.

How to Remove Duplicate Notes

One of the most basic, yet useful, MIDI editing commands is "Remove Duplicate Notes." It is very easy to accidentally hit a note on a MIDI keyboard twice—you might hit it tentatively just before it should be played and then again more positively on the beat or a little late. If a note starts within the first 25% of the duration of a note of the same pitch that is already sounding (or within an eighth-note, whichever is shorter), it is considered a duplicate and is combined with the previous note. If it starts later than that, the first note is shortened so that it ends at the same tick at which the new one starts. To use this command, make an Edit selection that includes the duplicate notes, then choose "Remove Duplicate Notes" from the Event menu.

How to Deal with Stuck Notes

A typical problem that comes up time and time again when you are programming MIDI instruments is "stuck notes." Typically, this happens when you change "patch" before the patch you are playing has received a MIDI Note Off command. If this happens, you can use the "All MIDI Notes Off" command from the Event menu or press Command-Shift-Period (.) if you are using a Mac or Control-Shift-Period (.) if you are using Windows. This will send out a series of messages to all the MIDI devices attached to your interface instructing them to turn off any stuck notes.

Mirrored MIDI Editing

Mirrored MIDI Editing lets you edit MIDI regions and have your edits apply to every MIDI region of the same name. This can be particularly useful when editing looped MIDI regions.

To enable Mirrored MIDI Editing, you can select the "Mirror MIDI Editing" item from the Options menu. To disable this feature, deselect this menu item. You can also use the Mirrored MIDI Editing button in the Toolbar section of the Edit window to "toggle" this feature on or off.

To warn you that your edit is being applied to more than one region, when you make an edit with Mirrored MIDI Editing enabled, the Mirrored MIDI Editing button will blink Red, just once—so keep an eye out for this.

Tip

It is a good idea to disable Mirrored MIDI Editing when you are not using it, as it is all too easy to inadvertently edit a region not realizing that this is editing other regions of the same name when you don't want to do this.

Selecting and Playing Notes with the Mini-Keyboard

You can use the mini-keyboard at the left of each MIDI or Instrument track in the Edit window to select and play notes by clicking on its keys with the mouse—with any Edit tool selected. When you click a key on the mini-keyboard, all the notes at that pitch on the track immediately become selected and ready for editing.

If you don't want the notes to be selected—because you just want to hear a note at that pitch play—hold the Command and Option keys (Mac) or the Control and Alt keys (Windows) while you click on the mini-keyboard.

If you want to select and play a range of notes, just click a key and drag up or down on the mini-keyboard, and if you want to lengthen or shorten the range of notes selected, you can Shift-click on the mini-keyboard.

> **Tip**
>
> If you want to deselect notes or select notes that are not adjacent to one another, hold the Command key (Mac) or Control key (Windows) and click on the notes.

Do to All

Two extremely useful keyboard commands let you apply changes to all tracks or to whichever tracks you have selected. For example, maybe you want to set all the track outputs to the same destination (Do to All). Or maybe you want to set all *selected* track outputs to the same destination (Do to All Selected).

To "Do to All," press and hold Option-click (Mac) or Alt-click (Windows) as you make your change. To "Do to All Selected," press and hold Option-Shift-click (Mac) or Alt-Shift-click (Windows).

Special Cut, Copy, Paste, and Clear Commands

The Edit menu commands, Cut Special, Copy Special, Paste Special, and Clear Special can be very useful for editing MIDI controller data. These each have three submenu selections two of which let you edit either all the MIDI controller data or just the MIDI pan data. (The third submenu command only works with audio plug-in automation.) Conveniently, these commands work whether the data is showing in the track or not.

The Cut Special, Copy Special and Clear Special commands work identically. So the "All Automation" submenu selection cuts, copies, or clears all MIDI controller data and the "Pan Automation" submenu selection cuts, copies, or clears only MIDI pan data.

The Paste Special submenu commands require further explanation: The "Merge" submenu selection pastes MIDI controller data from the clipboard to the selection and merges it with any current MIDI controller data in the selection.

The "To Current Automation Type" submenu selection pastes MIDI controller data from the clipboard to the selection and changes this to the current MIDI controller data type. This lets you convert from any MIDI controller data type to any other MIDI controller data type. For example, you could copy MIDI volume data and paste this to MIDI pan.

The "Repeat to Fill Selection" submenu selection repeatedly pastes the MIDI controller data from the clipboard until it fills the selection—saving lots of time compared with manual pasting of selections. Simply cut or copy a MIDI region, then make an Edit selection and use this command to fill the selection. And if you have selected an area that is not an exact multiple of the copied region size, it will automatically trim the last copied region that is pasted so that it fits exactly.

Graphic Editing in the Edit Window

Pro Tools MIDI tracks can be edited graphically in the Edit window. Here, you can use the standard Pro Tools Trimmer tool to make notes shorter or longer and use the Grabber tool to move the pitch or position—or "draw" notes in using the Pencil tool. You can draw in or edit existing velocity, volume, pan, mute, pitch bend, aftertouch, and any continuous controller data, and the Pencil tool can be set to draw freehand or to automatically draw straight lines, triangles, squares, or randomly. The Pencil tool also lets you draw and trim MIDI note and controller data, and the Trim tool can trim MIDI note durations when a MIDI track is set to Velocity view.

It can be very handy at times to insert notes using the pencil tool instead of setting up an external keyboard. Just make sure that the MIDI track is in Notes view and select the Pencil tool at the top of the Edit window. To insert quarter notes on the beat, for example, set the Time Scale to Bars and Beats, then set the Edit mode to Grid, and the Grid value to quarter notes. As you move the Pencil tool vertically and horizontally within a MIDI track, the bar/beat/clock location and the pitch of the MIDI note are shown in the Cursor Location and Cursor Value displays just below the Main Counter in the Edit window toolbar. When you find the note and position you want, you can just click in the track to insert it. It's as simple as that!

To select notes for editing, you can either use the Grabber tool to drag a marquee around the notes or drag using the Selector tool across a range of notes. Once some notes are selected, you can drag these up or down to change the pitch using the Grabber or Pencil tools—while pressing the Shift key if you want to make sure that you don't inadvertently move the position of the notes in the bar. You can use the Trimmer tool or the Pencil tool to adjust the start and end points of the notes. If you set the MIDI track to Velocity view, you will see the attack velocities of the notes represented by "stalks." You can edit these using the Grabber tool.

Sometimes, you may simply need to edit one note. If you select a note using the Grabber or the Pencil, its attributes will be displayed in the Event Edit area to the right of the counters in the toolbar at the top of the Edit window. Here, you can type in new values for any of the displayed parameters. See the accompanying screenshot for more details.

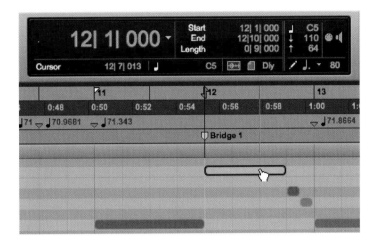

Figure 4.23 The selected MIDI note's attributes are displayed in the Event Edit area to the right of the Counters, ready for editing. The Main Counter shows the bar/beat/clock location of the start point of the selected MIDI note. The bar:beat:tick location of the cursor (12:7:013) and the pitch of the MIDI note that the cursor is pointing to (C5) are shown in the Cursor Location and Cursor Value displays just below the Main Counter.

Track Views

You can switch the Track View of any MIDI track in the Edit window to show MIDI Continuous Controller data. When Continuous Controller data is recorded onto MIDI tracks, it is displayed as a line graph with a series of editable breakpoints. These breakpoints are stepped to represent individual controller events, in contrast to the standard automation breakpoints, which are interruptions on a continuous line.

Figure 4.24 MIDI Continuous Controller data line graph with stepped breakpoints.

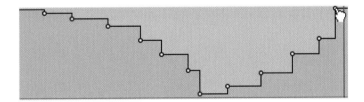

The Track View for Instrument tracks can either show MIDI CC data or it can show audio automation data, such as audio volume.

Figure 4.25 Audio Automation data vector graph with breakpoints on continuous line.

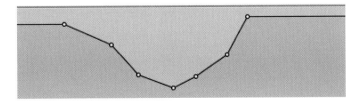

You can edit MIDI CC data directly in the Edit window according to which type of data you select using the pop-up Track View selector. Also, blocks, regions, and notes views, the Track View selector for MIDI tracks lets you choose velocity, MIDI volume, mute, pan, pitch bend, mono aftertouch, program change or sysex, and lets you access other controller views using the controllers submenu.

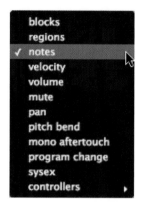

Figure 4.26 MIDI Track View selector.

The Track View selector for Instrument tracks additionally lets you display audio volume and volume trim, audio mute, pan left and right, and any send or plug-in automation data.

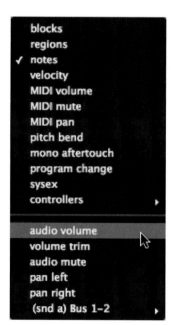

Figure 4.27 Instrument Track View selector.

For example, with MIDI volume selected in the Track View, you can use the Pencil tool to draw MIDI Volume data.

Figure 4.28
Drawing MIDI
Volume data in the
Edit window on a
MIDI track using
the Pencil tool.

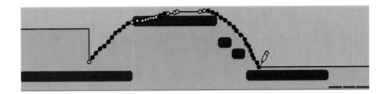

Automation Lanes

You can display the automation data more clearly in a separate lane below the notes display in the Edit window. Click on the disclosing arrow at the bottom left of the Track controls to show the automation lanes and choose the type of automation that you want to display using the pop-up selector provided. You can add more automation lanes using the small "+" icon, and you can remove these using the small "–" icon in the controls area at the left of each automation lane.

Figure 4.29
Revealing the
automation data
in a separate lane
below the note
data can make this
easier to edit.

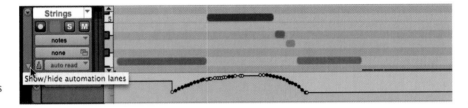

In Velocity View, you can edit each note's attack velocity using the Grabber tool by dragging its velocity stalk upwards or downward: the taller the stalk, the higher the velocity value. You can use velocities to create crescendos or to make accents within MIDI parts.

Figure 4.30
Velocity data
shown in an
Instrument track
overlaid on top of
the note data.

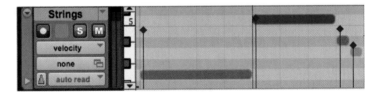

It can also be easier to edit the velocity data if you open a separate lane for this, but it does take up more space on your screen.

Figure 4.31
A lane opened
below the MIDI/
Instrument track
to allow Velocity
editing.

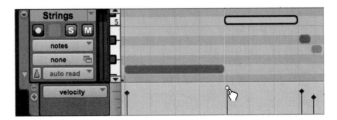

> **Note**
>
> MIDI controller #7 (volume) and #10 (pan) are treated by Pro Tools as automation data, so these controller events (along with Mutes) can be recorded and automated from the Mix window. Consequently, a MIDI or Instrument track's automation mode will affect how these events are played back and recorded. For example, if you suspend automation for a MIDI track, the MIDI volume, pan, and mute events will be suspended, but any other MIDI controller events will continue to operate. To avoid confusion, remember that Instrument tracks also support audio volume, pan, and mute along with MIDI volume, pan, and mute.

Note and Controller Chasing

New MIDI and Instrument tracks have Note Chasing enabled by default, but this can turned on or off individually for each MIDI or Instrument track from the pop-up Playlist selector menu.

Figure 4.32
Enabling Note
Chasing using the
Playlist selector
pop-up menu.

Note Chasing is necessary to allow sustained MIDI notes to be heard when you start playback at some point after the note's start time. If a long note started playing back at the start of the chorus at, say, bar 9 and lasted for 8 bars, you would not hear this note if you started playback half-way through this chorus section at bar 13—unless, of course, Note Chasing is enabled.

> **Note**
>
> You should disable Note Chasing when you are working with samplers that are playing loops because Note Chasing would trigger the loop to start playback from wherever you start the session playback—inevitably putting the loop out of time with the other tracks unless you start at correct the loop trigger location.

As far as continuous controller events and program changes are concerned, Pro Tools always chases these to make sure that controller values and any patches for MIDI devices are always set correctly.

Event List Editing

You can also view and edit MIDI data using the MIDI Event List, which is available from the Event menu. This lists MIDI events alphanumerically with letter names for the notes and numbers for the locations, the On and Off velocities, and the note lengths. In this window, you can select, copy, paste, or delete events and you can edit values by typing directly in the list. A pop-up selector at the top left of the window lets you select which MIDI or Instrument track to display in the event list.

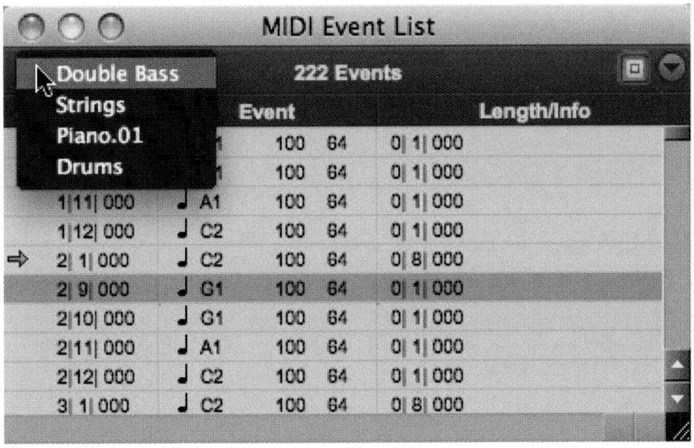

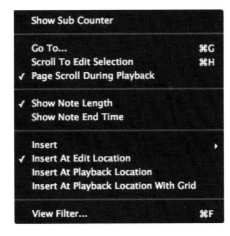

Figure 4.33 Data in the MIDI Event List window.

A second pop-up at the top right of the window has various commands that let you customize the MIDI Event List, make it scroll the way you want it to, or choose where events are to be inserted. You can also choose what to display in the list using the View Filter. A submenu, labeled Insert, lets you insert any MIDI event (except SysEx) into the list.

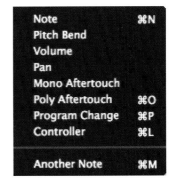

Figure 4.34 MIDI Event List Options pop-up selector.

Figure 4.35 MIDI Event List Insert submenu.

> ## Tip
>
> When the MIDI Event List is the front-most window, you can simply press Command-N (Mac) or Control-N (Windows) to insert a note, and use the left/right arrows on your computer keyboard to move between the Event Entry fields. To insert a Controller Event, press Command-L (Mac) or Control-L (Windows) and to insert a Program Change Event, press Command-P (Mac) or Control-P (Windows). After inserting any of these events, you can insert another similar event by pressing Command-M (Mac) or Control-M (Windows).

> ## Note
>
> Pro Tools has no list editor for audio event data—a major omission in my opinion.

Quantize to Grid Command

The Quantize to Grid command adjusts the placement of selected audio and MIDI regions (but not individual MIDI notes or individual Elastic Audio Events) so that their start points (or sync points, if present) align precisely to the nearest Grid boundary.

The way this works is that you choose a Grid value using the controls in the Edit window toolbar, use the Selector or Time Grabber to select the region or regions (on one or more tracks) that you want to quantize, then choose the "Quantize to Grid" command from the Region menu or press Command (Mac) or Control (Windows) and "0" on the numeric keypad. The Region start times (or sync points if these are different) will then be automatically aligned with the nearest Grid lines.

Figure 4.36 Part of the Regions menu showing the Quantize to Grid command.

With MIDI regions, only the region start times or sync points are quantized: all the MIDI data within the regions (such as notes) is moved equally, thereby retaining its rhythmic relationships. Similarly, with Elastic Audio regions, only the region start times or sync points are quantized: all Elastic Audio Events within the regions are moved equally, thereby retaining their rhythmic relationships.

Note

The Event Operation Quantize features can be applied to MIDI notes and Elastic Audio events individually in addition to audio region start times (or sync points).

MIDI and Instrument Controls and Track Heights

Each MIDI or Instrument track has a set of controls that includes the Track Name, buttons for Record Enable, Solo, Mute, and Patch Select, a Track View selector, a Playlist selector, a Track Timebase selector, an Automation Mode selector, and a Track Height selector.

To change the track heights for MIDI or Instrument tracks, you can click on the mini-keyboard at the left of the Track display in the Edit window and select the height you prefer.

Note

When the MIDI or Instrument track is in Notes or Velocity view, you will need to Control-click on the mini-keyboard to access the track heights pop-up.

It can be useful to view just one particular note pitch within a MIDI or Instrument track, such as those that play a single MIDI note on a sampler, or when you have split the different instruments of a drum kit onto individual MIDI tracks. Pro Tools provides a Single Note height for this purpose.

You can also access the Track Height selector from the Track Options pop-up. When the track height is set to small, mini, or micro, the Track Options pop-up lets you access the Track Timebase and Track View options as well because the track controls area becomes too small to provide space for these other pop-up selectors in their normal positions.

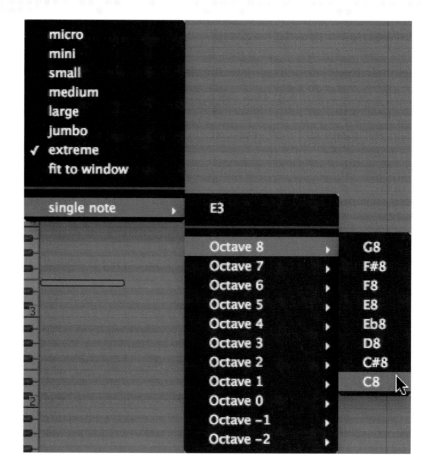

Figure 4.37 MIDI Track Height pop-up showing the available options (micro to extreme), including the single note options.

Figure 4.38 Single Note height.

Figure 4.39 Track Options selector.

Figure 4.40 Track Options pop-up.

In small view, there are three very small buttons provided to let you access the Track View, Patch Select, and Automation Mode selectors, but even these disappear in mini and micro views.

Figure 4.41 Small Track View showing three tiny buttons to the right of the Mute button for the Track View, Patch Select, and Automation Mode pop-up selectors.

MIDI Real-Time Properties

A relatively recent feature that users of Cubase and Logic will particularly appreciate is the MIDI Real-Time Properties. The "properties" that we are talking about here are Quantize, Duration, Velocity, Transpose, and Delay, all of which are attributes of MIDI notes that can be changed by selecting the notes and using the appropriate command from the Event menu. Delay is slightly different in that this can be achieved in a number of ways, one of which is by entering a MIDI Track Offset from the Event menu.

Using the Event menu commands changes the MIDI data, whereas using the MIDI Real-Time Properties feature just alters the data as it is being played back—in real time—leaving the original data unaltered. Why would you want to do this? Because sometimes you may want to experiment with different settings before you are sure which you want to commit to. You can even change the Real-Time Properties parameters while the session is playing back so that you can immediately hear the effect this will have.

When you have made up your mind, there is a command in the Track menu, "Write MIDI Real-Time Properties" that lets you write these properties to the track—permanently changing the data in the track.

Real-Time Properties for the whole track can either be adjusted in the Real-Time Properties View in the Edit window or by using the Real-Time Properties window that you can access from the Event menu.

You can also apply MIDI Real-Time Properties to individual regions by selecting the region or regions and opening the MIDI Real-Time Properties dialog from the Event menu. Region-based Real-Time Properties can only be applied using the Real-Time Properties window—not in the Edit window.

You will see the letter "T" displayed in the upper right-hand corner of the track's MIDI regions and in the Event List whenever Real-Time Properties are being applied to a track. Similarly, you will see the letter "R" displayed in the upper right-hand corner of the region and in the Event List whenever Real-Time Properties are being applied to a region.

Figure 4.42 Edit window showing the MIDI Real-Time Properties for a track. This track has just one region, identified with a "T" in the top right-hand corner to indicate that track-based Real-Time properties are being applied.

Figure 4.43 The MIDI Real-Time Properties dialog. Below this, you can see a MIDI region with the letter "R" displayed in the top right-hand corner to indicate that region-based MIDI Real-Time Properties are being applied to this region.

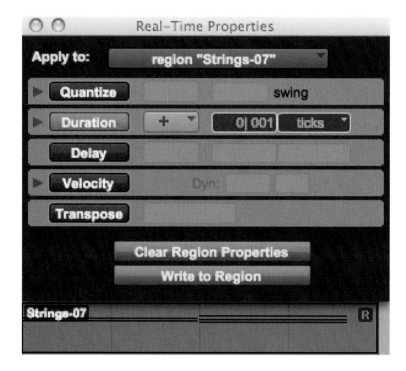

Note

If you wish, you can permanently change the data in the selected region, or regions, by clicking on the "Write To Region" button in the MIDI Real-Time Properties dialog.

The good thing about using MIDI Real-Time Properties is that you can change any of the parameters without having to commit to your changes until you are ready to do this. Also, if you are applying these to a track, you can keep these properties conveniently displayed in the Edit window if you wish, which can be a big help when you are working out what to do with your MIDI tracks. And when you are sure that you have the settings you want, you can write these settings to the selected tracks or regions.

Exporting MIDI Tracks

There will be times when you want to export MIDI tracks (or Instrument tracks containing MIDI data) from a Pro Tools session. For example, if you have recorded SysEx data containing patch information for your MIDI synthesizers or other external MIDI devices into Pro Tools tracks, you may wish to store this data separately for use with Pro Tools or other software in the future, or you may have developed useful MIDI sequences that you want to be able to build libraries of to use with Pro Tools or other software in the future. If you have recorded an arrangement using MIDI tracks in Pro Tools, you may wish to transfer this to other software, such as Digital Performer or Logic Pro, to work on the MIDI arrangement further, or you may just want to transfer individual tracks to a notation software application so that you can print out parts for musicians to play.

For these purposes, Pro Tools lets you export your Pro Tools session, or one or more MIDI or Instrument tracks, as a Type 1 (multitrack) or Type 0 (merged) Standard MIDI file. The MIDI file contains any meter and tempo information and any key signatures used in your session, along with track names and markers.

Figure 4.44
Export MIDI
dialog.

To export a selected track or a group of selected tracks, choose the "Export MIDI" command from the File menu in Pro Tools. The Export MIDI dialog allows you to choose the MIDI file format, the location reference (e.g., Session Start) and has a check box that you can tick if you want to apply Real-Time Properties to the exported data.

To export a single MIDI or Instrument track as a MIDI file you can Control-click (Mac) or Right-click (Windows) on the name of the track you want to export and choose "Export MIDI..." from the pop-up menu that appears. This shortcut also works with a group of selected tracks if you Control-click (Mac) or Right-click (Windows) on the name of any one of them.

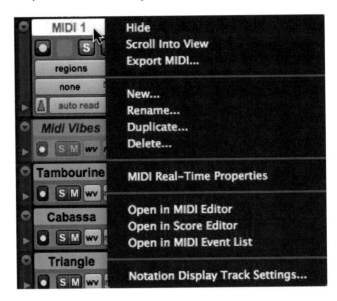

Figure 4.45 Track Name pop-up.

The MIDI Editors

Pro Tools provides separate MIDI Editor windows that can be better to use for detailed MIDI editing than the Edit window. The notes are displayed in a clearer way, for example, with velocity, controller, and automation displayed in separate lanes below the notes display.

To open a MIDI Editor, select any MIDI region and choose MIDI Editor from the Window menu. You can also open a MIDI Editor for any MIDI region by double-clicking on the region when the preference for this is set appropriately in the MIDI Preferences window. You can also use the keyboard command Control-equals (=) on the Mac or Start-equals (=) on Windows.

Figure 4.46 Part of the MIDI Preferences window showing what may be opened when double-clicking a MIDI Region.

Tracks and Groups in the MIDI Editors

You can view the Tracks and Groups Lists at the left of any MIDI Editor window and use this to reveal or hide Tracks and to enable or disable Groups, as necessary. To open the Tracks and Groups List, you can either click on the small arrow at the bottom left of the MIDI Editor window or use the Toolbar menu at the top right of the MIDI Editor window.

Figure 4.47 The MIDI Editor Window showing all the Edit Tools and the Grid/Nudge Display in the Toolbar and with the Tracks and Groups Lists open at the left.

By default, MIDI, Instrument, and Auxiliary Input tracks are all shown in the Tracks List. If you prefer, you can show only the MIDI tracks or only the Instrument tracks using the options in the Tracks List pop-up menu.

Figure 4.48
Using the Tracks List pop-up menu to choose what to show in the list.

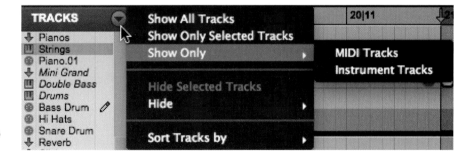

How the MIDI Editors Work

Each MIDI Editor window has a small red "target" button near the top right of the window. If you click this, it will turn gray. Then, if you open another MIDI Editor window, whether this is for the same MIDI region on the same track or for a different MIDI region on the same or on another track, this will be opened in a separate window. Otherwise, with the target button showing red, the new region will open in place of the existing MIDI Editor window.

If you choose to display multiple tracks in a single MIDI Editor window, the MIDI notes from the different tracks will be superimposed in the window.

Velocity and other continuous controller data for MIDI and Instrument tracks, and automation data for Instrument and Auxiliary Input tracks, are displayed in Controller lanes below the Notes pane in the MIDI Editor windows.

Velocities for MIDI notes are superimposed in the single Velocity lane, whereas other automation and controller data are shown using individual lanes for each type of automation or controller.

Using the MIDI Editor to Insert Notes

As with the main Edit window and the Score Editor, you can use the Pencil tool to insert notes onto any track in the MIDI Editor. However, because notes for multiple tracks can be displayed simultaneously in a single MIDI Editor window, Pro Tools lets you choose which of these to edit, preventing you from inadvertently making changes to the others. In the Tracks List, there is a column at the right that displays a pencil icon for whichever track is currently "pencil-enabled" for editing. If you only have one track showing in the MIDI Editor, that track will automatically be pencil-enabled. If you have multiple tracks, you can choose which to enable.

If the Tracks List is not open, the quickest way to open it is to select the track you want to edit using the Pencil-Enabled Track pop-up selector in the Toolbar. This pop-up selector lets you choose from whichever Tracks are currently being shown in the MIDI Editor window.

Figure 4.49
Pencil-Enabled
Track pop-up
selector.

If the Tracks List is open at the left of the window, you can simply click in the rightmost column to pencil-enable any of the Tracks that are currently showing and you can also show or hide any tracks by clicking in the leftmost column. If you want to be able to edit more than one track, you can Shift-click in the Pencil column to enable a contiguous selection of tracks. If any of the tracks in this range are not already being shown in the MIDI Editor, this action will cause them to be shown as well as pencil-enabled.

If the tracks you want to pencil-enable are not next to each other in the list, you can Command-click (Mac) or Control-click (Windows) in the Pencil column to enable additional tracks. And if you apply this command to tracks that are not currently shown in the Notes pane, they will be both shown and pencil-enabled. If you want to both show and pencil-enable *all* tracks in the Tracks List, the shortcut is to Option-click (Mac) or Alt-click (Windows).

To insert new notes, choose the Pencil tool from the Toolbar and click in the Notes pane at any time and pitch location. This action will insert a MIDI note on the pencil-enabled track. This will have the Default Note On Velocity unless you change this velocity first by entering a new value for Note On Velocity using the MIDI Editing Controls display. The note entered will also use the Default Note Duration shown in the MIDI Editing Controls, unless you change this first.

Figure 4.50
Pencil-enabling a track, ready to edit notes on this rack in the Notes pane.

Automation Lanes in MIDI Editor Windows

MIDI Editor windows also let you edit velocities, MIDI controller data, and automation for all shown MIDI, Instrument, and Auxiliary Input tracks in lanes under the Notes pane. To show or hide lanes under the Notes pane, just click the Show/Hide Lanes button. This is a small revealing arrow to the left of the keyboard at the bottom left of the Notes pane.

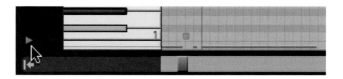

Figure 4.51
Using the Show/
Hide Lanes button.

If there are multiple MIDI and Instrument tracks showing in the Notes pane, the velocities for notes on separate tracks will all be shown together in a single lane, superimposed just like notes in the Notes pane.

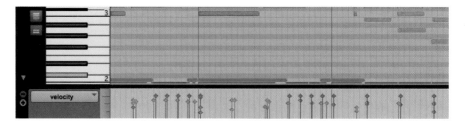

Figure 4.52
Superimposed
Notes and
Controllers.

However, all other Controller lanes are grouped by automation and controller type and provide individual lanes for each shown track. To add or remove lanes, you can use the "+" or "−" buttons at the left of the lane.

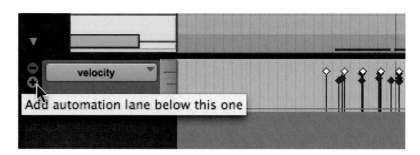

Figure 4.53
Using the Add
Lane button.

If the vertical zoom for the controller lanes is set to the minimum size, you won't be able to see the Add Lane button. In this case, if you use the pop-up Lane View selector, the options to add or remove a lane will be shown there.

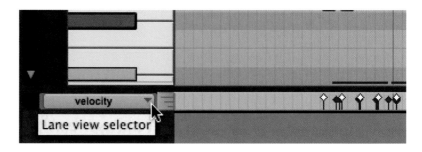

Figure 4.54
Using the pop-up
Lane View selector.

You can use the pop-up Lane View selector whenever you like to choose the Automation or Controller type you want to display. MIDI tracks let you choose from the standard list of MIDI controllers, whereas Instrument tracks also let you choose from the available audio automation types.

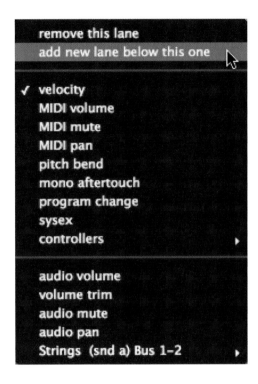

Figure 4.55
Pop-up Lane View selector showing available MIDI controller and audio automation types.

If you choose MIDI volume, for example, you can click using the Grabber tool to insert breakpoints, dragging these into position or moving existing ones to new positions, or draw in breakpoints using the Pencil tool to create and edit automation data.

Figure 4.56
Editing MIDI volume controller data using the lanes in the MIDI Editor window.

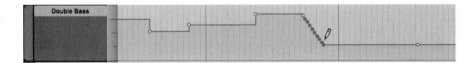

The MIDI Editor Toolbar

Running along the top of the MIDI Editor window you will find a toolbar that is similar to the Edit Window toolbar, but with a number of differences.

Mute and Solo

At the far left, there are two small buttons for Solo and Mute. The Solo button lets you solo all the tracks currently displayed in the MIDI Editor window, whereas the Mute button lets you mute all the tracks currently displayed in the MIDI Editor window.

Figure 4.57 Solo and Mute buttons engaged and applying to all tracks currently displayed in the MIDI Editor window—normally you would engage one or other of these.

If you have more than one track showing in the MIDI Editor window, engaging the Solo or Mute button in the MIDI Editor solos or mutes all these tracks. If, on the other hand, you solo or mute some, but not all, of the tracks shown in the MIDI Editor window using Solo and Mute buttons in the Edit or Mix windows, the appearance of the Solo and Mute buttons updates to indicate mixed states. Mixed solo and mute states are said to occur when one or more tracks shown in the MIDI Editor window, but not all, are soloed or muted.

Figure 4.58 Mute status indicates a mixed state, revealing that not all the selected tracks are muted.

Figure 4.59 Solo status indicates a mixed state, revealing that not all the selected tracks are soloed.

Notation View

To the right of the Solo and Mute buttons, there is a button that lets you switch the note display to show the MIDI notes as music notation on a stave. Each MIDI and Instrument track is displayed independently in Notation view, with one track per staff. Also, notation is displayed as a continuous timeline and not in page view as in the Score Editor window. You can edit notes in Notation view just like you can in the Score Editor (as explained in the next section), but with the advantage of having access to Velocity, Controller, and Automation lanes.

> **Note**
>
> In Notation view, the Trimmer tool functions as the Grabber tool when you hover this above a note and functions as the Note Selector tool when it is not over a note. Also, the Edit mode is automatically set to Grid and cannot be changed.

Figure 4.60
Notation display
enabled in the
MIDI Editor
window.

At the top left of the Notation view, just below the rulers, there is a Double Bar button. If you enable this button, a double bar line will be displayed at the end of the last MIDI region or event in the session. When this is disabled, there will be a number of empty bars at the end of the last event in the session. You may want additional empty bars to be available so that you can manually enter new notes after the last MIDI region or event in the session if you have not yet completed your composition. In this case, you should disable the Double Bar button, and you can specify the number of empty bars using the "Additional Empty Bars in the Score Editor" setting in the MIDI Preferences page that you can access from the Setup menu.

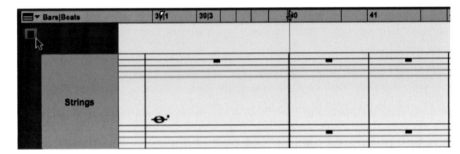

Figure 4.61
Clicking on the
Double Bar button
in Notation view.

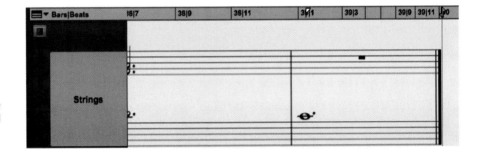

Figure 4.62
A double bar line
indicates the end
of the piece of
music.

The Edit Tools

The button to the right of the Notation button is a pop-up selector that defaults to the Zoomer tool.

Figure 4.63 Edit Tool Pop-up Selector— defaulted to the Zoomer Tool.

If you click and hold this button, a pop-up selector opens from which you can choose an Edit tool or the Smart tool.

Figure 4.64 Edit tools pop-up selector.

You can expand this section of the Toolbar to show these Edit tools as a row of buttons, just like in the Edit window. To open the Expanded Edit tools, click on the pop-up Toolbar menu at the top right of MIDI Editor window.

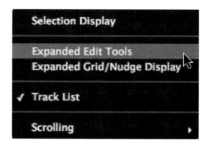

Figure 4.65 Toolbar menu.

The Edit tools in the MIDI Editor window let you work with MIDI data in the same ways as the tools in the Edit window.

> ## Note
> The Edit tool settings can be set differently for each window when you have more than one MIDI Editor window open.

Figure 4.66 The Expanded MIDI Editor tools.

MIDI Editing Controls

To the right of the Edit tools, you will find a set of MIDI Editing controls that work exactly like the MIDI Editing controls in the Edit window.

Figure 4.67 The MIDI Editing controls.

The leftmost of these controls is the Track Edit pop-up Selector that lets you choose which of the visible tracks to edit. Avid calls this "pencil-enabled" and warns that you can only insert MIDI notes manually in "pencil-enabled" tracks.

Next along is the Default Note Duration selector that you will use to choose note durations when you are manually inserting MIDI notes. A useful choice here is to follow the durations set using the Grid.

Figure 4.68 Default Note Duration pop-up selector.

The numerical entry field with the number 95 set by default is the default Note On Velocity setting for MIDI notes that you manually insert. You can type any number between 1 and 127 here. The combined MIDI socket and loudspeaker icon is actually a button that, when enabled, causes MIDI notes to play when you insert them using the Pencil tool, click them with the Grabber tool, Tab to them, or select them with the Note Selector tool.

Additional Toolbar Buttons, Controls, and Displays

To the right of the MIDI Editing controls there are two more large buttons. The first of these is the Link Timeline and Edit Selection button that should be familiar to you from the Edit window. If you disable this, you can make independent Timeline and Edit selections. The second of these buttons lets you enable or disable Mirrored MIDI Editing. This lets you edit one MIDI region and have your edits apply to every copy of that same MIDI region.

Moving further along the Toolbar you will find the standard Edit Mode buttons, which operate in exactly the same way as those in the Edit window. There is also a Cursor Location display that shows the exact location that the cursor is pointing at in the MIDI Editor window and also displays the pitch that the cursor is currently positioned at.

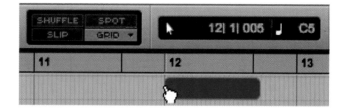

Figure 4.69 The Edit Mode buttons and the Cursor Location Display.

To the right of the Cursor Location Display, an Edit Selection Display can be added to the Toolbar by selecting this from the Toolbar menu. Here, for example, you can see the exact Bar:Beat:Tick location of a selected note, along with its MIDI Note Name and Note On Velocity.

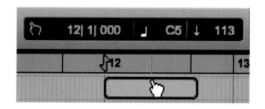

Figure 4.70 The optional Edit Selection Display.

The final group of controls for Grid and Nudge can be displayed in an either compact or expanded format. By default, the compact form has a pop-up selector that lets you choose whether to set the Grid or the Nudge values.

Figure 4.71 The compact version of the Grid and Nudge controls, with the cursor pointing to the pop-up Grid/Nudge Toggle selector that lets you choose which to control.

From the Toolbar menu, you can select the expanded format, which displays the Grid and Nudge controls separately. To the right of these controls, you can see the Target button, which is colored red when it is enabled, and the Toolbar menu pop-up selector.

Figure 4.72 Expanded Grid and Nudge controls with the enabled Target button and the Toolbar menu pop-up selector to the right of these.

The Target button works the same way that the Target button works in Plug-in, Output, and Send windows, allowing you to have untargeted windows open at the same time. When you untarget any of these windows by clicking on its red target button, this button becomes grayed-out to confirm that it is untargeted.

The Targeted MIDI Editor window (you can only target one MIDI Editor window at a time) also synchronizes its Timeline location view to the Timeline location view in the Edit window. Consequently, if you make or change an Edit selection in the Edit window, this same Edit selection will be made or changed in the Targeted MIDI Editor window.

The Toolbar menu has a couple of items that have not been mentioned yet: you can open the Track List at the left of the MIDI Editor window by selecting the Track List item here, and a submenu is provided to let you choose from various Scrolling options for the MIDI Editor window. The Scrolling options operate independently from those chosen for the Edit window and can be set independently for each MIDI Editor window. There is also an option to have the MIDI Editor window follow the Scrolling settings that you have made for the Edit window.

> **Note**
>
> You can rearrange the positions of the controls and displays in the toolbar, just as you can in the Edit window: Command-click (Mac) or Control-click (Windows) and simply drag the controls or displays that you want to a new location in the Toolbar.

The MIDI Editor Zoom Controls

There is a Horizontal Zoom control for the MIDI Editor window, consisting of a pair of "+" and "–" buttons, tucked away at the bottom right corner of the window, just to the right of the scrollbar that runs along the bottom of the window. This controls the vertical zoom for both the Notes pane and the Controller lanes.

There is a Vertical Zoom button for the Notes pane located in the upper right corner of the MIDI Editor window, just above the right-hand scroll bar. Clicking the top part of this button, this will increase the vertical zoom and clicking the lower half will decrease the vertical zoom for MIDI notes.

A second pair of "+" and "–" buttons located just beneath the scrollbar at the right-hand side of the window acts as a Vertical Zoom control for the Automation and Controller lanes that run below the Notes pane.

Timebase and Conductor Rulers

You can choose which Timebase and Conductor rulers to view in each MIDI Editor window independently of each other and independently of the main Edit window, using the pop-up selector just underneath the Solo and Mute buttons. You can also view different time locations from the main Edit window in one or more MIDI Editor windows, and these can be set to different zoom levels.

Figure 4.73 MIDI Editor window showing Timebase and Conductor rulers.

The Superimposed Notes View

The MIDI Editor windows allow you to view and edit MIDI notes from different MIDI and Instrument tracks at the same time, all together, superimposed in the same window. To distinguish notes from different tracks, these can be color-coded by Track or by Velocity using two buttons at the top left of the MIDI Editor window, just below the rulers.

The first button lets you enable the "Color Code MIDI Notes By Track" feature. When this is enabled, tracks in the MIDI Editor are temporarily assigned one of sixteen fixed colors, in the order they appear in the track list.

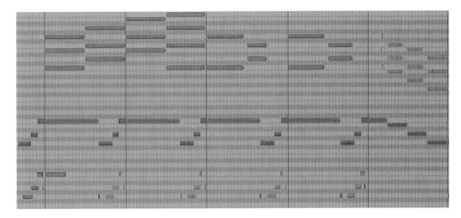

Figure 4.74 MIDI Editor color coded by Track.

The second button lets you enable the "Color Code MIDI notes by Velocity" feature. When this is enabled, all the MIDI notes are all displayed in red, but the color saturation varies from light red for low velocities to dark red for high velocities, based on the Note On velocities.

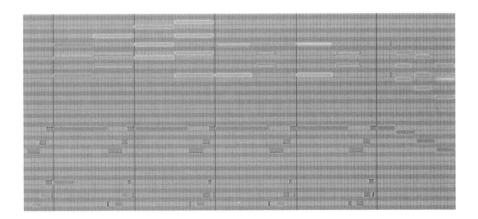

Figure 4.75 Color coded by velocity.

When neither button is enabled, MIDI notes are drawn using the same colors they are in the Edit window, using the color coding options that you can set in the Track Color Coding settings in the Display Preferences.

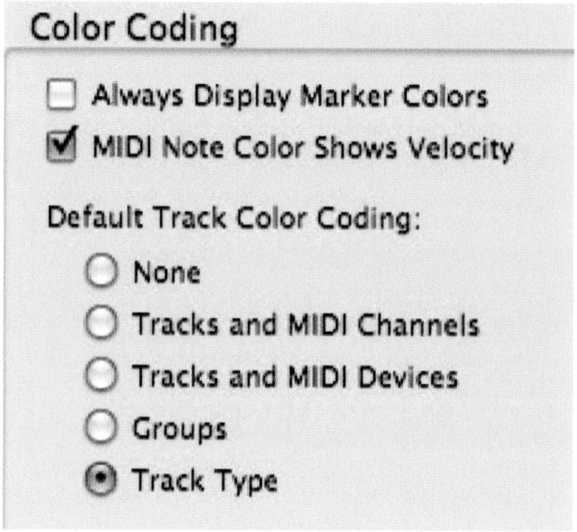

Figure 4.76
Display
Preferences: Color
Coding options.

The Score Editor

One of the best new features for music production in Pro Tools 8 is the Score Editor. Anyone who reads and writes conventional music notation will immediately realize the benefit of this. Just ask the guitar player to read the chords onscreen or print off a copy of the chord sheet so he or she can learn the part, or sketch out a bass part using MIDI, then print this out and ask a bass player to turn this into a "proper" bass part. Put some strings or brass parts together—you get the idea…

It won't take you long to learn how to use the Score Editor either—it has one main window with a simple tool palette and just a couple of dialog boxes to let you adjust things like the page layout for printing. The tools are the same ones that you will recognize from the Edit window—the Zoomer, the Trimmer, the Selector, the Grabber, and the Pencil—so if you know Pro Tools already, you will know how to use these. Everything is very intuitive. If you see something you want to change, such as the Title or a Clef or a Time Signature, just double-click it and it will usually open up a dialog window to let you change it immediately.

Using the Score Editor

To get started with the Score Editor, you could record some piano or synthesizer pad chords, or import a MIDI file, so that you have some music to work with.

A quick way to get some music into the Score Editor before you have learned how to use its specialized tools is to enter the chords for your music using the Chords ruler.

After you have entered the chords, create a new MIDI track (if there is not one already there) and open the Score Editor. Here, you will see the chords laid out above the default music staves. But there will probably be lots of empty bars after the music has finished. To fix this, just click the Double Barline button that you will find among the tools at the top of the Score Editor window and this will create a double barline marking at the end of your score—and the unused bars will disappear from the screen.

Figure 4.77
Entering a Double
Barline at the end
of the music.

If you have used a lot of chords, and especially if you are not a guitarist, you probably won't want to clutter up your score with lots of guitar chord diagrams. To get rid of these, open the Score Setup dialog by double-clicking on the Title in the Score Editor window. Here you can untick Chord Diagrams in the Display section so that these will not be displayed. You can also enter the title and the composer credit in the Information section.

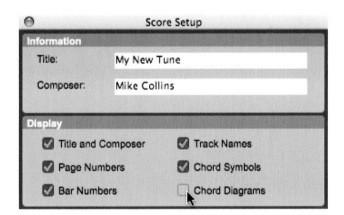

Figure 4.78 Part
of the Score Setup
dialog showing the
Information and
Display options.

At the bottom right of the Score Editor you will find the Score zoomer arrows. You can use these to size up the score so that you can see just what you need to see in the window.

Figure 4.79
Score Zoomer.

Take a look at the accompanying screenshot to see how the Score Editor window might look when you have set everything up this way, with the piano chords visible above the piano stave.

Figure 4.80
Piano chords revealed in the Score Editor.

The vertical blue line is called the Cursor Location indicator. This indicates where the music will start from when you press the Spacebar or click Play on the Transport. If you want to move this, you can either type another location into the Main counter or just use the mouse to grab it and move it to wherever you would like to start playback from.

When you play back the music, a separate playback cursor appears in the Score Editor window and moves through the score as the session plays back. This is a little thinner and is black, not blue, so you won't get them confused.

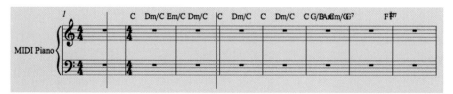

Figure 4.81
Playback Cursor visible near the end of bar 3.

> **Note**
>
> You will need plenty of screen space if you want to see lots of the Score Editor onscreen, especially if you are using lots of instruments. A second monitor, or even a 30 in. monitor, would be good to have. If you are using a smaller monitor, you may not be able to see the whole score, but you can resize the Score Editor to show just one or two lines of the score and set it to scroll during playback.

Score Editor Pop-up Menu

In the top right-hand corner of the Score Editor window, there is a pop-up menu that lets you choose whether or not to show the Selection Display and the Expanded Edit Tools in the toolbar that runs along the top of the window.

Figure 4.82
Opening the Score
Editor Menu.

It also lets you open the Tracks List at the left of the window (which allows you to show or hide your MIDI tracks in the Score Editor) and lets you set the Scrolling options for the Score Editor window.

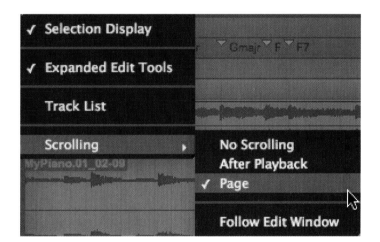

Figure 4.83 The
Score Editor Menu.

As you add more MIDI or Instrument tracks, these will show up in the Score Editor using the default piano stave. To change these to single staves, you can either select "Notation Display Track Settings" from the Tracks List pop-up menu in the Score Editor or just Double-click a Clef on a Staff.

The Tracks List pop-up menu lets you show or hide selected tracks or all tracks, and lets you access the Score Editor's two main dialog windows for "Notation Display Track Settings" and "Score Setup."

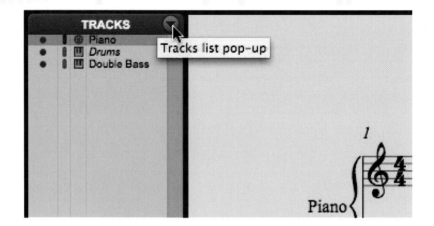

Figure 4.84
Tracks List pop-up
menu.

Figure 4.85
Tracks List pop-up
menu.

Setting the Clef

In the "Notation Display Track Settings" window, you will find a pop-up selector that lets you choose between the Grand Staff that you use for piano, harp, organ, and other keyboard instruments, or the Treble, Bass, Alto, and Tenor clefs used for instruments such as bass, drums, percussion, vocals, strings, or brass.

Figure 4.86 Part
of the Notation
Display Track
Settings dialog
showing the clef
settings.

Drum Notation

Music for kit drums is typically written using the bass clef and the standard five-line musical staff. The hi-hat or cymbal notes are usually written above the top line of the staff, at musical pitch B, written with stemmed X's. You will sometimes see the letters OP to indicate open and CL to indicate closed written over the hi-hat. If there is no indication, the hi-hat remains closed. The small tom-tom is indicated by a note-head placed in the top space, corresponding to the musical pitch G; the snare is in the space below at musical pitch E; the large tom-tom is below this in the space at musical pitch C; with the bass drum in the space below this at musical pitch A. The hi-hat may also be found in the space below the lowest line at musical pitch F, especially if the top line of X's is being used for the ride cymbal. There can be variations to this scheme, with drums written on the lines instead of in the spaces, or with two separate bass clef staves for clarity—usually there will be notes with the music that explain all this.

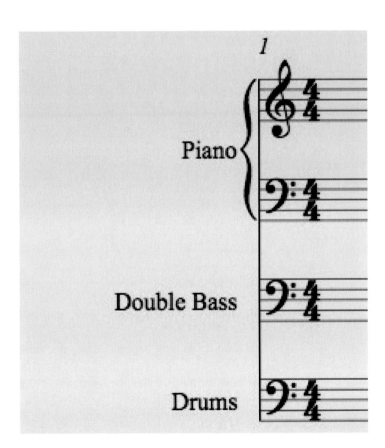

Figure 4.87 Clefs set correctly for Piano, Drums, and Double Bass.

Hiding Tracks

To hide MIDI or Instrument tracks that you don't need to see at any particular time, open the Tracks List and click on the small black dot to the left of any track you want to hide. This will immediately disappear from the Score Editor window.

Figure 4.88
Hiding tracks.

Using the Notation Tools

To insert a note, you just choose the pencil tool, select a note duration using the pop-up selector to the right of the pencil tool, then click at the pitch and timing location within a bar where you want the note to appear. Rests will automatically be inserted elsewhere in the bar.

Whenever you insert or edit a note in the Score Editor, this note will also appear on (or be changed in) a MIDI or Instrument track in both the Edit and MIDI Editor windows—it's all the same MIDI data.

Figure 4.89
Selecting a MIDI Note Duration.

If you want to hear the MIDI notes play when you click on these in the score, click on the speaker icon to the right of the MIDI Note Duration selector and leave this highlighted. Then, when you click on any note in the Score Editor, you will hear it play back—assuming that the MIDI or Instrument track that it is based on is set up to play back correctly.

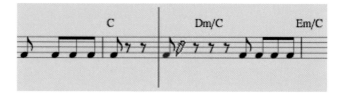

Figure 4.90
Using the Pencil tool to insert notes.

Trimming Notes

If you want to lengthen or shorten a note, just use the Trimmer tool as you would with a MIDI note or an audio region.

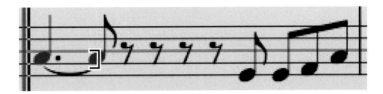

Figure 4.91
Trimming a note.

If you point at a selected note on the staff, dragging to the left will make it shorter and dragging to the right will lengthen it.

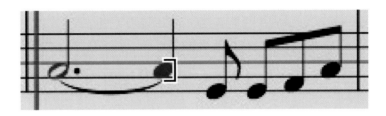

Figure 4.92
Using the Trimmer tool to lengthen a note.

When you get used to working this way, it won't take you too long to enter a complete bass part for a song.

Figure 4.93 The Double Bass part for my new arrangement laid out in the Score Editor.

Copying Notes

Copying notes is easy enough. For example, I decided to copy some of the bass notes in my first arrangement to create a matching drum part. First, I used the Note Selector tool to select the notes in the score, dragging the cursor across the notes to highlight them in blue.

Figure 4.94
Selecting notes using the Note Selector Tool.

With the notes selected, I used the standard Copy and Paste commands from the computer keyboard to copy these to an Instrument track that I had set up for the drums.

Figure 4.95
Notes copied to the Drums track.

> **Note**
>
> Before pasting the notes at a specific location, you need to make sure that the Cursor Location indicator is positioned at the correct location within the bar. You can just drag the blue Cursor Location indicator to any location you like within a bar. Alternatively, you can just type the location into the Main Counter.

Moving Notes

You can use the Grabber tool to move selected notes to any other time or pitch location on the same track. With the Grabber tool selected, simply click and drag to move the notes.

You can also move notes to another time or pitch location on the same track using the Pencil tool. This changes into a pointing finger that you can use to grab and move notes when you point the mouse at existing notes with the Pencil tool selected.

Figure 4.96
Moving Notes
using the Grabber
tool.

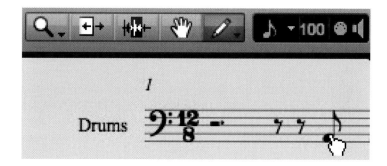

Figure 4.97
Using the Pencil
tool to move a
note.

Transposing Instruments

A "transposing instrument" is a musical instrument for which music is written at a different pitch than the pitch that you will hear when the instrument is played. There are two common reasons why this is done: either to make the music easier to read or to make the fingerings on the instrument easier to learn and play. And sometimes both of these reasons apply.

If an instrument has a very high or very low range, the music is typically written either an octave higher or lower than it actually sounds to reduce the use of ledger lines. Examples sounding an octave higher include the celeste and the xylophone, piccolo, and tin whistle. Examples sounding an octave lower include the guitar, bass guitar, double bass, and bass flute. Some instruments with extremely high or low ranges, such as the glockenspiel, use a two-octave transposition. Instruments that "transpose at the octave" in this way don't play in a different key from concert pitch instruments; instead they just sound an octave higher or lower than written. For example, music for the double bass is written on the bass clef, one octave higher than concert pitch, and music for the guitar is written on the treble clef, one octave higher than concert pitch. Music for the piccolo, on the other hand, is written on the treble clef, one octave lower than concert pitch.

The instruments in families of different sized instruments such as the saxophone, clarinet, and flute, have differing ranges of notes that they can play, and these instruments sound lower as they get larger. So that a player can play any member of a particular instrument family using the same fingerings, these instruments are transposed, based on their range, such that each written note is fingered the same way on each instrument. Transposing instruments of this type are often referred to as being in a certain "key," such as the horn in F. This instrument "key" determines which pitch will sound when the player plays a note written as C. So, for example, if you play a written C on the French horn (horn in F), it will sound the note F and if you play a written C on the Bb clarinet, you will hear a Bb note. All the other notes on a transposing instrument will, correspondingly, sound at different pitches than written.

Here are some common examples: Instruments in Bb include the Soprano Saxophone, the Bb Clarinet, the Trumpet, and the Flugelhorn and sound a major second lower than written. Instruments in G include the Alto Flute and sound a perfect fourth lower than written. Instruments in F include the Cor Anglais and the (French) Horn and sound a perfect fifth lower than written. Instruments in Eb include the Alto Saxophone and sound a major sixth below what is written. Low-bass instruments in Bb include the Tenor Saxophone and the Bb Bass Clarinet (which is sometimes written in treble clef) and sound an octave and a major second below what is written. Bass instruments in Eb include the Baritone Saxophone and sound an octave and a major sixth below what is written. Very low-bass instruments in Bb include the Bass Saxophone and sound two octaves and a major second below what is written.

How the Score Editor Handles Transposing Instruments

The Score Editor lets you take account of transposing instruments by allowing you to display notes for any track with a transposition that does not affect the MIDI notes—it just displays the pitches differently in the Score Editor.

You can set these display-only transpositions using the "Notation Display Track Settings" window. The Key pop-up selector lets you choose the key signature that will be displayed and the Octave slider lets you transpose up or down by one or two octaves in either direction.

With a Bass Clarinet track recorded using a MIDI instrument to sound at the correct pitch, for instance, you would set the Key pop-up to B-flat and set the Octave slider to +1. Then, if you print out the Bass Clarinet part, a musician will be able to read the part easily and the pitch will still sound correctly—a major ninth below where the part is written.

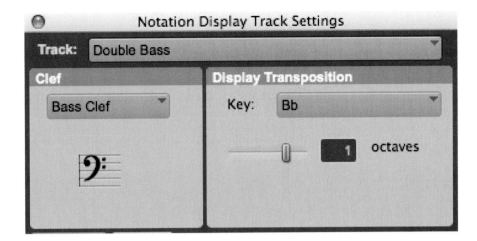

Figure 4.98
Notation Display
Transposition.

To get music notation into the Score Editor, you can either write the notes directly into the Score Editor or record them onto a MIDI or Instrument track using step entry or using a MIDI keyboard. You don't have to assign the output of the MIDI track to any instrument if all you want to do is to see the music notation, but most people will probably find it very convenient to play the part using a virtual or real MIDI instrument so that they can hear how this sounds while they are building up the arrangement.

> **Tip**
>
> If you only want to get notes into the Score Editor so that you can print these out for a musician to read, you might as well go ahead and write these at the transposed pitches in the transposed key. Then, when you print this out, the musician will just play what he sees while the sound from his instrument will be at the correct pitch.

Preparing the Score for a Musician to Read

When you record a musical part with a particular MIDI or virtual instrument, especially if there is an element of "swing" or a lot of complexity in the rhythms, this can look much too difficult for a musician to read while they are trying to play this.

Composers and arrangers often simplify the music notation to some extent when they are preparing parts for musicians to play. Musicians understand this and will interpret straight notes that are intended to be "swung" according to the directions of the leader, musical director, arranger, or producer—or according to how they feel the music should be performed. They will make decisions about exactly how long to sustain the notes and what accents to give the notes in the same way—guided by whoever is directing their performance or according to their own experience.

Accordingly, the Score Editor allows you to choose a "Display Quantization" so that the notes in the score are easier for a human to read, while the underlying MIDI data can use whatever quantization is appropriate so that the sounds produced by MIDI or virtual instruments are correct.

Display Quantization

The "Notation Display Track Settings" window has two tabs that let you set global "Display Quantization" and other attributes for all the tracks or individual "Display Quantization" attributes for individual tracks.

Individual Attributes

By default, all the tracks use the global attributes, so if you click on the Attributes tab, you will see that "Follow Globals" is selected and the various attributes are grayed out.

If you want to set these attributes individually for a particular track, choose the Attributes tab, make sure this track is selected in the Track pop-up at the top of the Notation Display Track Settings window, de-select "Follow Globals," then go ahead and set the attributes that you want for this track.

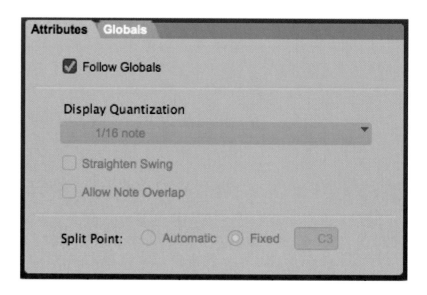

Figure 4.99
Notation Display
Attributes set to
"Follow Globals"

Global Attributes

If you choose the Globals tab, you can set the various attributes that you want to apply to all the tracks that are not individually set up.

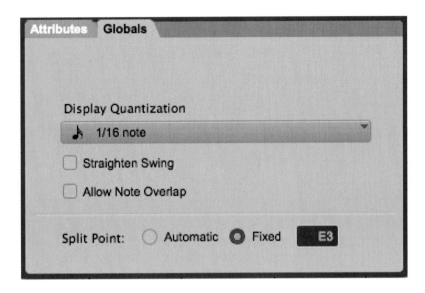

Figure 4.100
Notation Display
Global Attributes.

The Attributes

The main attribute that you will set is the Display Quantization, using the pop-up selector provided.

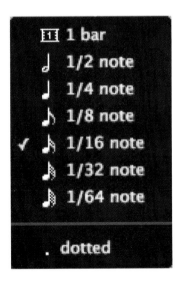

Figure 4.101
Display
Quantization
pop-up selector.

If you choose thirty-second note quantization, for example, the thirty-second note subdivisions can be difficult to read at a glance when sight-reading.

Figure 4.102
1/32 note quantization.

If you set the display quantization to sixteenth notes, it doesn't affect the MIDI data in the track, but it does make the music easier to read—especially for a session musician or band member who is feeling the pressure of the "red light" in the studio.

Figure 4.103
Sixteenth note quantization.

A check box for "Straighten Swing" lets you set "swung" notes to be displayed as straight notation. If you select this box, a run of "swung" eighth notes would be displayed as straight eighth notes, for example.

Sometimes, two notes that start at different times overlap. If this happens, the first note will be truncated when the second note begins. If the "Allow Note Overlap" check box is selected, the Score Editor will display the full length of any overlapping notes using tied notes.

You can also set a Split Point to divide notes between the treble and bass clefs. When the Clef for the selected track is set to Grand Staff, the selected Split Point setting determines the pitch at which the notes are placed in either the upper or lower staff of the Grand Staff.

If you click in the button marked "Automatic," the notes will be split between the upper and lower staves of the Grand Staff based on logical note groupings. If you click in the button marked "Fixed," you can use the associated alphanumeric field to specify a fixed pitch at which to split notes between the upper and lower staves of the Grand Staff.

Printing the Score

When you have prepared all the tracks in the Score Editor to look the way that they should, there are still a couple of things that you may need to attend to in the Score Setup window before you print out the individual tracks or the complete score.

Display Options, Spacing, and Layout

In the Score Setup window you can choose whether to display the Title and Composer credits, the Track names, the page numbers, the bar numbers, the chord symbols, or the chord diagrams. As I mentioned earlier, you would not want to clutter up the pages with guitar chord diagrams unless you really do need these. Even more importantly, you can adjust the spacing between the various elements on the page and adjust the layout to make sure that everything is as easy to read as possible.

To open the Score Setup window, either choose Score Setup from the File menu or select Score Setup from the Tracks menu in the Score Editor window, or you can simply double-click on the Title.

In the Spacing section, you can enter the spacing you want to use between the staves and between systems of staves. You can also specify the distance to leave below the Title and Composer credits and the first staff of the score and below the chord symbols and diagrams and the top stave of each system.

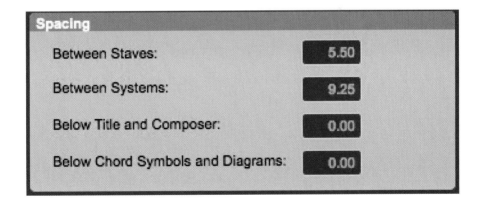

Figure 4.104
Score Setup
Spacing options.

The Layout section has a pop-up selector to let you choose the page size: A4, Tabloid, Legal, or Letter. You can also choose whether to use Portrait (tall and not as wide) or Landscape (wide but not as tall) page orientation. You can also specify the Staff Size and the Page Margins in inches or millimeters.

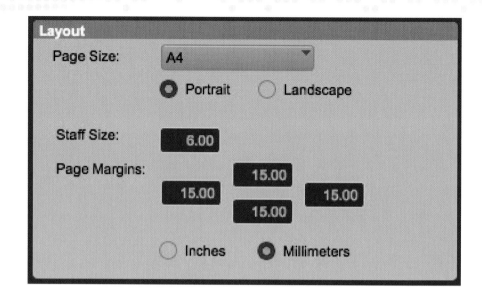

Figure 4.105
Score Setup
Layout options.

What You See Is What You Get!

When you have everything else set up and ready to print your score, all that is left is to choose whether to print the whole score or individual tracks. What gets printed is what you see in the Score Editor, so all you need to do is to hide any tracks that you don't want to print. If you want to print individual tracks, hide all the other tracks and just show one track at a time, print this, then show the next track and print that, and so on.

To print the score you can either choose "Print Score" from the File Menu in Pro Tools or press Command-P (Mac) or Control-P (Windows). Your computer's standard Print dialog will appear and you may choose further options here, or just click Print (Mac) or OK (Windows) to print your score.

Exporting to Sibelius

Pro Tools also lets you export the score from your session as a Sibelius (.sib) file. You can then open your score in Sibelius, edit the notation further, and print the score and parts from Sibelius. To export a score from Pro Tools, choose "Export:To Sibelius" from the File Menu, select a destination, and click Save.

You can also send the score directly from your Pro Tools session to Sibelius if you have Sibelius installed on your computer. To do this, you can either Right-click in the Score Editor and choose "Send to Sibelius" or choose "Send to Sibelius" from the File menu. Pro Tools will export all the Instrument and MIDI tracks shown in the Score Editor to Sibelius as a .sib file and will launch Sibelius if it is installed on your computer. As with printing, all you need to do is to hide any tracks that you don't want to export.

Summary

Pro Tools offers a rich set of features for working with MIDI data. If you have MIDI hardware in your studio, such as synthesizers, samplers, drum machines, keyboards, or other controllers, or control surfaces to provide hands-on control for virtual instruments and plug-ins, then Pro Tools provides the tools you need to record all your MIDI data. You can choose your method—graphical, Event list or Score—and use the powerful editing features that are provided in the Event Operations and MIDI Real-Time Properties windows.

The separate MIDI Editor and Score Editor windows provide even greater ease-of-use than the main Edit window for many types of editing, but the main Edit window allows you to access everything you need without cluttering up your screen unnecessarily. The MIDI Editor has the advantage that you can view MIDI notes from different tracks superimposed in the same window if you wish—which can be very useful when working on complex musical arrangements. And the Score Editor has the advantage that it is very easy to use and clear to read.

Time spent getting your MIDI equipment organized and set up to work with Pro Tools and preparing templates that can be reused will be time well spent. If you work out any offsets needed to tighten up the timing aspects of your MIDI gear, this will make a big difference to the results you will be able to achieve.

You should always record MIDI and virtual instruments as audio as soon in your workflow as possible so that you can then reclaim the DSP resources used to create these for use elsewhere in your session. This makes it easier to transfer projects to other software and ensures that you can always hear these parts even if you no longer have access to the hardware or software instruments.

Pro Tools started out as an audio recording, editing, and mixing environment. It now offers an extremely capable MIDI recording, editing, and mixing environment as well.

In This Chapter

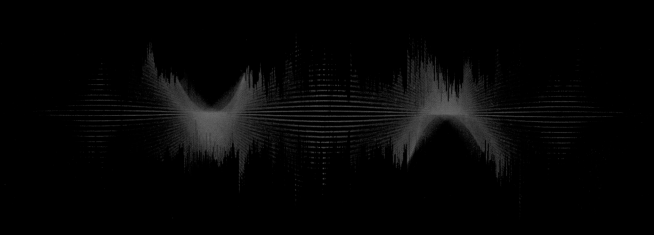

Recording

Introduction

A complete recording system includes microphones and microphone preamplifiers; optional signal processors such as equalizers and compressors; mixing features for multiple microphones; a recording capability with enough channels/tracks; playback facilities for existing music or for a click to play along with; playback facilities for the newly recorded music; a cueing system with headphone amplifiers and headphones for the musicians to hear the click tracks or music playback; and a loudspeaker monitoring system with a talkback system and controls so that the recording engineer, producer, arranger, or others involved in the recording can hear everything they need to and communicate with the musicians.

Pro Tools systems do not include monitor speakers or microphones, and professional Avid Pro Tools hardware interfaces, with the exception of the HD Omni, don't incorporate microphone preamplifiers—they typically have line-level analog or digital inputs—so you will need to use additional Avid or third-party equipment in most recording situations. Avid does offer the PRE, which is a high-quality 8-channel remotely controllable microphone preamplifier unit. However, many recording engineers and producers prefer to use established designs from third-party manufacturers. So, most studios will want to add at least a couple of high-quality microphone preamplifiers to the equipment list, along with at least one or two (or even a cupboard full of) high-quality studio microphones if they plan to record vocals, guitars, or other acoustic instruments. Another option, especially for larger studios, is to use a professional mixing console from Neve, SSL, API, or other quality manufacturers. These mixing consoles usually have extremely high quality microphone preamplifiers and plenty of signal processors, such as equalizers or compressors, built-in. A recent trend from third-party manufacturers is to produce mixing consoles that are designed especially to interface with Pro Tools systems, such as the SSL AWS 900 series, which also functions as a Digital Audio Workstation (DAW) controller, or the small format API 1608, which provides 16 vintage-style analog mic pres and 16 equalizers. An external mixer also provides latency-free monitoring while recording; usually has additional connections for playback equipment; and typically provides volume controls, dim, mono, and mute switches, talkback features, and switching for control room and studio monitor systems. Headphone amplifiers with suitable sets of headphones, such as the BeyerDynamic DT100 models, will be needed for the musicians while they record. The recording engineer or producer will be more interested in the super high-quality BeyerDynamic DT48 headphones, which are much more expensive to buy, but which produce sound quality similar to professional studio monitors. For the engineer and producer in the control room, Avid offers its Reference Monitor Series of powered nearfield monitors, which nicely complement Pro Tools systems. Of course, there are plenty of alternatives from companies such as ATC or PMC at the high end, Genelec or KRK in the mid-ground, and more affordable models from Mackie and others at entry-level.

Preparing Pro Tools for Recording

If you are just recording one singer or musician, singing or playing without accompaniment and without a click, then all you need to do is to route a microphone through a suitable mic pre via the A/D converters in your Pro Tools interface onto one audio track. Ask the performer to sing or play the loudest that they expect they will during the recording, and then set the gain on the microphone preamplifier to avoid distortion while keeping the level high enough to maintain a good signal-to-noise ratio. Record-enable the track, press Play, and Record in the Transport window, and go ahead with your recording.

While you are learning to use Pro Tools or testing a newly installed system for the first time, it's a good idea to plug a microphone in and try this out first. As soon as you want to get a little more ambitious, recording with more microphones, it will pay you to make a number of other preparations. For example, it's always a good idea to check the Pro Tools I/O setup window and enter helpful names for the Inputs, Outputs, and Busses. By default, a newly recorded audio file and its region are named based on the name of the Track onto which it is recorded. So, you should also name the track or tracks that you are recording onto with names that will make sense when you look through the recorded "takes" to find these later on in the Regions list.

You should make sure that you have set the tempo and meter appropriately and set up a click track for the musicians to play too. If they will be playing along to existing material, you may need to create and adjust one or more headphone mixes and route these into the studio for the musicians. It should never be underestimated how important the headphone mixes can be if you want to get the best results from the musicians. It is always a good idea for the engineer or producer to go out into the studio and listen to the headphones personally to make sure that everything is OK. Often, the musicians may not be able to successfully describe what is wrong with the sound or to recognize faults that will be immediately apparent to the engineer or producer.

If you have an external mixer or a DAW controller or monitor controller with talkback facilities, you will be able to talk to the artist from the control room. If not, you should set up a microphone in your control room, route this into an Auxiliary Input in your Pro Tools mixer, and route the output to the musician's headphones—not forgetting to mute this when you are not using it. It can also be a good idea to set up an extra microphone in your recording room, set to omni, so it "listens" to everything, and route this into an Auxiliary channel in your Pro Tools mixer. This way, the musicians can communicate with you even if their instrument or vocal mics are muted.

Disk Space for Recording

Do you have enough disk space to record for as long as you will need to? Pro Tools lets you check how much drive space is available. The Disk Usage window, available from the Window menu, shows the available drive space for each drive connected to your system as text and as a gauge display.

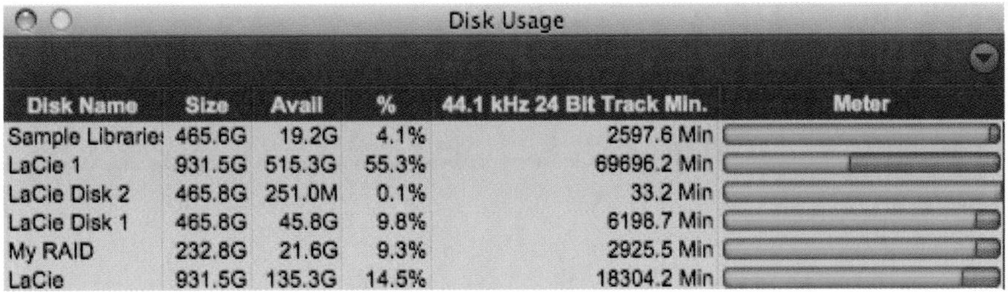

Figure 5.1
Disk Usage
window.

Disk Name	Size	Avail	%	44.1 kHz 24 Bit Track Min.	Meter
Sample Librarie:	465.6G	19.2G	4.1%	2597.6 Min	
LaCie 1	931.5G	515.3G	55.3%	69696.2 Min	
LaCie Disk 2	465.8G	251.0M	0.1%	33.2 Min	
LaCie Disk 1	465.8G	45.8G	9.8%	6198.7 Min	
My RAID	232.8G	21.6G	9.3%	2925.5 Min	
LaCie	931.5G	135.3G	14.5%	18304.2 Min	

You can limit the amount of your available hard disk space that is allocated for recording using the Record Allocation preference settings in the Operation Preferences window. This defaults to the Open-Ended "Use All Available Space" setting, but you can choose to limit this to a specified number of minutes if you want to make sure that you don't actually use all the available space. It is never a good idea to use up all the available space on your disk drives, as this leaves no room for moving things around or for any system-related requirements. The rule of thumb is to keep 10% of your disk drives free at all times.

Figure 5.2
Setting the Record
Allocation in
the Operation
Preferences
window.

Open-Ended Record Allocation:

⦿ Use All Available Space

◯ Limit to: 60   minutes

> **Tip**
>
> If you do choose to allocate all available space, Pro Tools may take a little longer to begin recording. To avoid this delay, you can put Pro Tools into Record Pause mode before you start to record.

> **Note**
>
> In general, the "Use All Available Space" preference makes hard drives work harder. In addition to record and punch lag times, many systems see better overall recording performance when the "Open Ended Record Allocation" setting is limited.

Setting Up Headphone Mixes

In addition to the main mix that you will use to monitor the recordings in the control room, you may also need to create headphone or "cue" mixes for the musicians and vocalists. The best way to do this is to use Sends to route audio from each track to a pair of available outputs on your audio interface, and route these in turn to a suitable headphone-monitoring amplifier. You can also set up a Master Fader to control the overall level for your cue mix. And if you need different headphone cue mixes, you can use additional Sends, Master Faders, interface outputs, and headphone amplifiers.

To set up a cue mix for all the tracks in your session, you can press Option (Mac) or Alt (Windows) and choose an available Send to insert this on every track. You probably won't want to insert a cue Send on every track, though there will usually be some tracks that it makes no sense to route into the headphones. So, more usefully, there is a shortcut to insert Sends on selected tracks only: first you select all the tracks on which you wish to insert Sends for your cue mix, and then press the Shift key at the same time as holding the Option (Mac) or Alt key (Windows) and choose an available Send.

> **Tip**
>
> To select a group of tracks, Shift-click on the track names in the Mix or Edit window. To select nonadjacent tracks, hold the Command (Mac) or Control key (Windows) and click on the nonadjacent tracks.

Often, you will be using output paths 1 and 2 from your audio interface to route the main mix from Pro Tools to your monitors. If this is the case, output paths 3 and 4 may be free to use for your cue mix, so you would route the Sends to these outputs. You would then need to route this output pair from your hardware interface into a suitable headphone monitoring amplifier into which you would plug the headphones.

One way to set up a cue mix is as follows: set each of the Sends to unity gain by Option-clicking (Mac) or Alt-clicking (Windows) on the send fader in each Send window. Also, set each Send to post-fader and enable the Follow Main Pan (FMP) button on each send so that the cue mix will have the same panning as the main mix. This lets you use the same mix for the cue as for your main mix. Of course, you can always increase or decrease the level of any individual track in the cue mix using an individual Send fader if the singer or the guitarist wants to hear "more me."

Another way to set up a cue mix is to hold the Option (Mac) or Alt key (Windows) and use the "Copy to Send" command to copy the current fader settings in the Pro Tools Mix window to the sends. You will find this command in the Automation submenu in the Edit menu (Pro Tools HD and Pro Tools with Complete Production Toolkit 2 only).

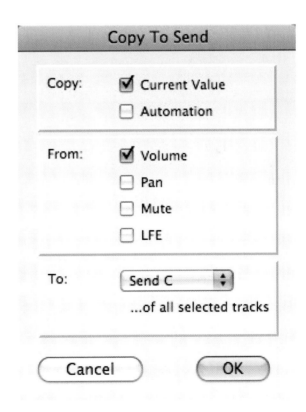

Figure 5.3 Send window in post-fader mode set to unity gain with Follow Main Pan selected.

Figure 5.4 Copy To Send dialog.

If you Command-click (Mac) or Control-click (Windows) on the Send selector for any track, this will change the Sends view to show the individual assignments for level, pan, mute, and pre/post fader.

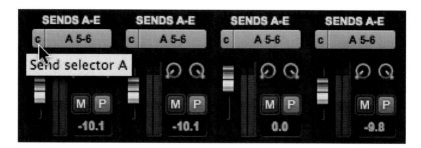

Figure 5.5 Clicking on the Send selector to change the Sends view. The Sends view is showing the individual assignments for Send C, which is being routed to interface outputs 5 and 6 in this example. Notice that the pre-fader buttons are highlighted in blue to indicate that these are enabled.

You can Option-click (Mac) or Alt-click (Windows) on the Pre/Post button for any Send to set all the Pre–Post buttons to pre fader (these will be highlighted in blue when selected) so that the Send levels will not be affected by any changes you make to the main mix fader levels. You can also tweak the individual send levels to suit the musicians without affecting the main mix.

If you want to change the Sends view so that you see a different individual send or so that you see all five send assignments again, Command-click (Mac) or Control-click (Windows) on the Send selector for any track and choose any individual Send or choose Assignments to see them all.

Figure 5.6
Choose Assignments from the Send selector pop-up to show all five send assignments at the same time or choose any individual Send to show individual assignments for the chosen Send.

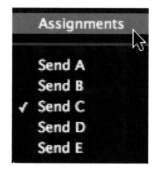

To control the overall level for the cue mix, create a new stereo Master Fader and set this to control the output paths that you are using for the cue mix, in this example, outputs 3 and 4.

Tip

If you need additional cue mixes, for the drummer or for other groups of musicians, you can set up additional Sends and route these via additional Master Faders to other available outputs on your audio interface in the same way.

Note

If you are working with Delay Compensation enabled, Avid recommends that you do not use any inserts on any Master Fader tracks that you are using to control cue mix levels. Also, you should avoid using inserts on any tracks that you are recording onto, although low latency inserts may not be a problem. Always check that latency is not a problem for the musicians and singers, and never forget that you are creating the cue mix to help the performers, not for your convenience!

Monitoring Latency

There will always be some delay, typically referred to as *latency*, between the audio coming into Pro Tools and the audio that is sent out from Pro Tools to your monitor speakers. At the very least, there will be a delay of a few samples in length due to the analog-to-digital conversion process on the way in and the digital-to-analog process on the way out. There may also be other latency delays caused by bussing signals around within the Pro Tools mixer or due to any processing of the audio signals within Pro Tools, which will take at least some small additional amounts of time. So, you will always hear a delayed version of the audio input through your monitor speakers when you are monitoring through Pro Tools.

Figure 5.7
Master Fader set up to control the overall level of a cue mix being sent to interface outputs 3 and 4.

You are unlikely to be able to hear the small amounts of latency caused by A/D and D/A conversions, but you may be able to hear delays caused by bussing signals internally or due to processing by plug-ins, and some plug-ins can cause quite long delays that you will have no trouble in hearing.

For example, RTAS plug-ins use the host processor in your computer, and monitoring latency will occur if you use any RTAS signal processing plug-ins on the tracks you are recording with. This is particularly noticeable when you are playing RTAS virtual instruments and monitoring the instrument's audio output, which will suffer from monitoring latency when the host processor receives the MIDI data and processes this via the plug-in to create the audio. If this delay is clearly audible, this can be very off-putting for a musician who is trying to overdub to an existing performance because the music already recorded will play back at the correct time while the new performance will be delayed with respect to this due to the monitoring latency.

It is possible to reduce the size of the hardware buffer to minimize latency delays. You can do this in the Playback Engine dialog that you can access from the Setup menu—setting the H/W Buffer Size to a smaller number of samples to reduce the latency. However, even at the smallest buffer size, there is still some latency. Also, reducing the buffer size limits the number of simultaneous audio tracks you can record without encountering performance errors. When you are recording, you will have to set the smallest possible buffer size to counteract latency, but when you have higher track counts with more plug-ins you will need to use larger buffer sizes.

> **Note**
>
> The lowest available H/W Buffer Size setting will be 64, 128, or 256 Samples, depending on the Number of Voices setting. For example, the 64-sample buffer size setting will only be available when the Number of Voices setting is 48 Voices (1 DSP) or 96 voices (2 DSP).

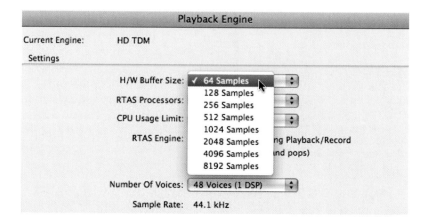

Figure 5.8
Setting the Hardware Buffer Size.

When Automatic Delay Compensation is enabled, this is applied to Sends and Inserts to compensate for mixer delays inherent in bussing. So, in most cases, you can now record in Pro Tools with Delay Compensation enabled to minimize latency problems.

Of course, you can avoid the latency problem altogether by monitoring the recording source using an external mixer, before it is routed to Pro Tools. Plug your microphones or instruments into the mixer and you will hear the sound directly through a connected monitor system, with no audible delay. Simultaneously, route the audio to your Pro Tools interface so that you can record it.

> **Tip**
>
> If you are monitoring through an external mixer while recording, don't forget to defeat the monitoring (pull the fader down or engage the mute button) for each track that you are recording in Pro Tools. Otherwise, you will hear a delayed version of the audio mixed in with the undelayed version.

Delay Compensation

When you are using plug-ins or hardware inserts and when you are routing audio within the Pro Tools mixer using sends and busses, delays can occur due to the time needed for the digital signal processing to take place. Pro Tools provides automatic Delay Compensation to compensate for these DSP delays. This Delay Compensation maintains phase coherent time alignment between tracks that have plug-ins with differing DSP delays, tracks with different mixing paths, tracks that are split-off and recombined within the mixer, and tracks with hardware inserts.

Delay Compensation should always be used while you are playing back and mixing in Pro Tools. It should also be used in most recording situations.

> **Note**
>
> When Delay Compensation is enabled, you will have to avoid using inserts on tracks that are being used for recording (or for controlling the levels of cue mixes) because Delay Compensation will create an unwanted time difference between the tracks playing back from disk and any new audio coming into Pro Tools.

To enable Delay Compensation, you can either select Delay Compensation from the Options menu or choose Short or Long Delay Compensation in the Playback Engine dialog that you can access from the setup menu. The Short Delay Compensation setting is the most efficient setting for Pro Tools|HD Accel systems, and if you are only using a few plug-ins that are not producing too much DSP delay, this should suffice. The Long Delay Compensation setting is intended for use on sessions with lots of plug-ins and large amounts of DSP delay.

Figure 5.9
Part of the Playback
Engine dialog
showing the Delay
Compensation
Engine settings.

> **Note**
>
> Pro Tools adds the exact amount of delay to each track necessary to make
> that particular track's delay equal to the total System Delay. The total sys-
> tem delay is the longest delay reported on a track, and any additional delay
> caused by mixer routing.

The total amount of delay due to inserts and mixer routing for the entire session is displayed in the Session Setup window that you can access from
the Setup menu. You can view the reported System Delay in the Session
Setup window to check whether or not you are close to exceeding the Delay
Compensation limit.

Figure 5.10
The total System
delay in samples
is displayed in
the Session Setup
window, accessible
from the Setup
Menu.

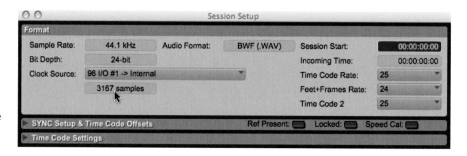

In the Operation Preferences page, the Delay Compensation Time Mode setting
lets you specify whether Delay values are displayed in samples or milliseconds.

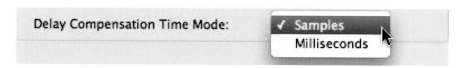

Figure 5.11 Delay Compensation time Mode pop-up selector in the Operation
Preferences.

There is a Delay Compensation status indicator in the Edit Window Toolbar that
is displayed when Delay Compensation is activated, but to see what is going on
with the delays on each track you should enable the Delay Compensation View

for tracks in the Mix Window. This view displays the total amount of delay due to TDM and/or RTAS plug-ins, and any hardware inserts, in the Delay Indicator "dly" field. You can apply a track delay offset using the User Offset "+/–" field, and you can view the total amount of delay that Pro Tools is applying to the track in the Track Compensation "cmp" field.

The alphanumeric characters in these fields are displayed in different colors to indicate status. For example, green is used to indicate that everything is working normally.

Orange is used to identify the track reporting the longest plug-in and hardware insert delay in the session.

Figure 5.12 The Structure Instrument track with sufficient delay compensation available.

Figure 5.13 An Aux track with Altiverb RTAS plug-in inserted has the longest delay in this session, so the delay parameters are shown in orange.

When the total amount of plug-in and hardware insert delay on a track exceeds the total amount of Delay Compensation available, the alphanumeric characters in the Delay Compensation view turn Red, and Delay Compensation is disabled on that track.

Figure 5.14 Delay Compensation View: the first track has no inserts; the Piano track has the Compressor/Limiter Dyn 3 TDM plug-in inserted; the Instrument track has Structure inserted and the delay values are shown in red, because with the Short Delay Compensation Engine selected there is not sufficient delay available to compensate for this amount of delay.

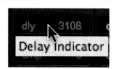

Figure 5.15 Delay indicator turned grey to show that the Delay Compensation has been bypassed.

If this takes place, check the Delay Compensation Engine setting in the Playback Engine and if this is set to Short, change the setting to Long. If the Delay Compensation Engine is already set to Long, you can manually compensate for the track delay by moving audio data on the track earlier by the amount of delay reported in the track's Delay indicator. In this case, you should bypass the reported delay for the track by Command-Control-clicking (Mac) or Start-Control-clicking (Windows) on the Delay indicator. The Delay Indicator (dly) field will turn Grey when Delay Compensation is bypassed.

While Delay Compensation is enabled, you can use the User Offset (+/−) field to adjust the timing "feel" of the track manually, adding to or subtracting from the amount of delay being applied by the automatic Delay Compensation. The User Offset field is displayed in White when a User Offset is being used on the track and in Grey when the User Offset feature is disabled on the track. To bypass the user offset delay, Command-Control-click (Mac) or Start-Control-click (Windows) the User Offset field.

The Track Compensation (cmp) field shows the amount of Delay Compensation being applied to each track. The Track Compensation field turns Grey when Delay Compensation for the track is bypassed, and no delay is being applied to the track.

When an audio track is record-enabled, Track Input Monitoring-enabled, DestructivePunch-enabled, or punched in, the track's Delay Compensation is automatically suspended to provide low-latency monitoring on the track outputs. If this is a problem, you can forcibly apply Delay Compensation, even when the track is record-enabled or Track-Input-enabled, in which case, the Track Compensation indicator turns Blue. To apply Delay Compensation to tracks on which Delay Compensation has been suspended, Command-Control-click (Mac) or Start-Control-click (Windows) on the Track Compensation indicator.

RTAS Plug-ins "Health" Warning (Pro Tools HD Only)

Using RTAS plug-ins can seriously deplete the number of available voices in your Pro Tools HD system. They also add to the latency in your system, so use with care! When you initially insert an RTAS plug-in on an Auxiliary Input or Master Fader or on an Instrument track that does not contain an instrument plug-in, and whenever you insert an RTAS plug-in after a TDM plug-in on any kind of track, this will use two additional voices per channel—one for input and one for output. And if the track is stereo, each channel will use two voices!

The bad news is that Pro Tools also uses another voice for the track if you use the external key side-chain of an RTAS plug-in on that track or if you select multiple track outputs for the track (it uses one additional voice for each additional output). And one voice is used for each channel when you select an AFL/PFL Path output in the Output tab of the I/O Setup dialog.

The good news is that adding further RTAS plug-ins on the same track does not use any additional voices—unless, that is, a TDM plug-in is inserted between two of these other RTAS plug-ins! Obviously, this is something you should, and can, avoid. Just remember that inserting TDM plug-ins between RTAS plug-ins on any kind of track will always cause unnecessary voice usage and may cause additional latency.

So, how does Pro Tools handle things when you mix TDM and RTAS plug-ins? Well, this is a little more complicated to explain. It depends on what mix of TDM and RTAS plug-ins you are using. Combining RTAS and TDM plug-ins on an Audio track, Auxiliary Input, or Master Fader, produces different results depending on the order in which they are inserted.

If the RTAS plug-ins are grouped together and inserted *before* the TDM plug-ins, no additional voices will be used, and no processing latency is added. With this configuration, the RTAS plug-ins will be bypassed when record is enabled or TrackInput monitoring is enabled for that track.

If the RTAS plug-ins are grouped together and inserted *after* the TDM plug-ins, each initial insert of an RTAS plug-in after a TDM plug-in on a track will cause processing latency and will use voices as described earlier. And, in this case, RTAS plug-ins will *not* be bypassed when Record Enable or TrackInput monitoring is enabled for that track.

As a consequence, there are two rules that you should observe when using combinations of TDM and RTAS plug-ins:

1. Always group plug-ins of the same type together and insert these before any TDM plug-ins to avoid additional voices being used and to minimize processing latency.
2. If you need to make sure that the RTAS plug-ins stay active when you record-enable a track or use TrackInput monitoring, then you must insert any TDM plug-ins before you insert the RTAS plug-ins.

Auto Input and Input Only Monitoring

Before you start recording, you also need to consider the monitoring mode and how to set the monitoring levels in the Pro Tools mixer. There are two different ways in which the input signals can be monitored while playing back, while recording, or with the transport stopped. These are the Auto Input and Input Only monitoring modes.

Let's look at the default Auto Input mode first. When playback is stopped, the track monitors whatever is coming into its audio input. Now, let's consider what happens when you are overdubbing and reach a punch-in point. In this case, you will normally have to hear whatever has been recorded up to the punch-in point. During the punch-in, you will hear the new audio that is coming in. When you reach the punch-out point you want the monitoring to switch back so that you can hear the existing track material again. So, this is exactly what Auto Input mode does.

Sometimes you don't want to hear the original material that was recorded, but you want to hear the incoming audio at all times. This is what Input Only monitoring does.

> ### Note
> These input monitoring options will be familiar to recording engineers who are accustomed to have input switching available on analog multitrack recorders.

You can toggle between Auto Input and Input Only monitoring by pressing Option-K (Mac) or Alt-K (Windows). Alternatively, you can select whichever of these monitoring modes is not currently in use by choosing Auto Input monitoring or Input Only monitoring from the Track menu to allow all record-enabled tracks to use these.

With Pro Tools, the Input Monitor Enabled Status indicator (in the Transport window) lights green when Input Only mode is enabled.

Figure 5.16 The cursor arrow points to the Input Monitor Enabled Status indicator.

With Pro Tools HD, the Input Monitor indicator lights green when one or more tracks have a TrackInput button enabled.

With Pro Tools HD, tracks are in Auto Input mode by default, and each mixer channel strip has a Track Input button that lets you switch individual audio tracks between Auto Input and Input Only monitoring modes. You can switch modes at any time, during playback, recording, while stopped, and even when a track is not record-enabled.

You can also press Shift-I to enable the TrackInput Monitor button for any track containing the Edit cursor or an Edit selection.

When you have record-enabled any tracks, the menu command "Set Record Tracks to Auto Input" becomes active in the Track menu. This lets you toggle the track input status of all record-enabled tracks between Auto Input and Input Only.

> **Note**
>
> When in Auto Input mode, the switch back to monitoring track material on punch-out is not instantaneous.

Figure 5.17
When Input Monitor is enabled on a track, the Input Status LED in the Transport window and the TrackInput Monitor buttons for each enabled track turn green.

There are several useful shortcuts for switching TrackInput monitoring: To toggle all tracks in the session, Option-click (Mac) or Alt-click (Windows) on any TrackInput Monitor button. To toggle all selected tracks in the session, Option-Shift-click (Mac) or Alt-Shift-click (Windows) on a selected track's TrackInput

> **Tip**
>
> If you have selected both Input Only and Record Enable on a track, Input Only monitoring will be disabled if you deselect Record Enable for the track. Sometimes you will not want this to happen, for example, if you are recording on a series of tracks, one at a time. To prevent this from happening, you can disable this behavior in the Operation Preferences by deselecting the "Disable "Input" When Disarming Track (in "Stop")" preference.

Monitor button. To toggle record-enabled tracks between Auto Input and Input Only monitoring, press Option-K (Mac) or Alt-K (Windows).

Setting Monitor Levels for Record and Playback

When you record-enable audio tracks, their volume faders turn bright red. This is to remind you that different fader levels can be set for any audio track during recording and playback. If you *deselect* "Link Record And Play Faders" in the Operation page of the Preferences window, you can set monitoring levels independently when Pro Tools is recording and when Pro Tools is playing back, for each audio track, and Pro Tools will remember the two different fader levels: one for when the track is record-enabled, and the other for when it is *not* record-enabled. When the Operation preference for "Link Record and Play Faders" is *selected*, the monitoring levels will be kept the same when Pro Tools is recording and playing back, keeping the mix consistent whether you are recording or listening.

Recording with or without Effects

You need to think carefully about whether to record any effects that you are adding to disk, or whether to use these effects purely for monitoring, perhaps to "vibe up" the performers while they are playing. Singers usually like to hear at least a little reverb, and guitar players often want to hear distortion or delay effects, for example. One problem with monitoring, but not recording, effects is that the musician will often play things that only make sense when hearing this particular effect and if you don't record this at the same time, you may never be able to recreate this again: you may not be able to get hold of the plug-in or the hardware that was used.

If you want to record the effects that you are adding to voices or instruments to disk, you have more choices to make. You may be adding the effects by bussing signals to and from external hardware. Or you may be routing the audio from the microphone preamplifiers into "outboard" effects before routing this into Pro Tools. It is also possible to use software plug-ins to create effects while you are recording.

Recording with Plug-in Effects

If you insert a plug-in on an Audio track and record audio into this track, the effect is not recorded to disk, it is only applied to the audio coming back from disk. If you want to record the effect to disk, you need to use an Auxiliary Input track. Connect your audio source to this input, insert one or more plug-in effects on the Auxiliary Input and route the output to the input of an Audio track. Then, you can use the Audio track to record the incoming audio together with the sound of the plug-in effects.

Figure 5.18
Recording with plug-in effects inserted on the Auxiliary Input. The output of this channel is routed to the input of an Audio track, which is used to record the incoming audio with the plug-in effects applied to this.

> **Note**
>
> The additional processing via the plug-in adds a further latency delay that can be considerable with certain plug-ins. This is why it is more usual to apply plug-in effects to your recordings after you have recorded the basic tracks.

Recording Audio

When you have set up the tracks you want to record onto, routed microphones into these, set the levels on the microphone preamplifiers, set up headphone mixes that the musicians are happy to work with, checked playback of any existing material, sorted out clicks, tempo, meter, key, chords, or whatever, and, ideally, set up markers for the sections of the music, set up any effects processing (either for the musicians to hear to "give them the vibe" or that will also be recorded to disk), and set your monitoring levels for the control room and for the headphone mixes, then you are close to getting started! But there are still more moves to make…

Record Enabling

Each track in Pro Tools has a Record Enable button that must be selected so that it lights up red before you put Pro Tools into Record mode using the Transport controls. You can record simultaneously to multiple tracks by record-enabling your choice of Audio, Instrument, or MIDI tracks. When you have clicked the record-enable button for the first track you want to set up, you can add more choices by Shift-clicking on more track record-enable buttons as you decide which you want to use.

Tip

If you enable the "Latch Record Enable Buttons" preference that you will find in the Operations section of the Preferences window, then you don't need to Shift-click the Record Enable buttons to enable additional tracks—just click each in turn that you want to be enabled. They will all stay enabled until deselected. And if you want to record-enable all the tracks in your session, if you are recording a band, for example, then Option-clicking (Macintosh) or Alt-clicking (Windows) will do this.

Pro Tools also provides keyboard shortcuts to let you record-enable, input-enable, solo, or mute any tracks that contain the Edit cursor or an Edit selection without using the mouse: simply hold down the Shift key and press R for Record, I for Input enable, S for Solo, or M for Mute to toggle these on or off. If you are working in the Edit window, it can be quicker to use these commands to input-enable and record-enable tracks.

When you have record-enabled all the tracks you want to record onto, click the Record button in the Transport to arm recording, and then click the Play button in the Transport to start recording. Alternatively, you can arm and start recording immediately by pressing 3 on the numeric keypad (when the Numeric Keypad mode is set to Transport) or by pressing the function key F12. Or you can simply press Command-Spacebar (Mac) or Control-Spacebar (Windows), which I always use.

Note

On Mac systems, to use F12 or Command-Spacebar for recording, the Mac "Spotlight" feature must be disabled or remapped.

If the take gets messed up for any reason, and you realize this before you stop recording, the fastest way to abort the recording is to press Command-Period (Mac) or Control-Period (Windows). This instantly aborts the recording without saving the file—saving you time if you realize that you have a useless take.

If you stop the recording, and then realize that it is not good, you can use the standard keyboard command, Command-Z (Mac) or Control-Z (Windows) to undo your last action, in this case, the recording that you just made.

Tip

If you are recording a large number of tracks, or playing back a large number of tracks while recording, Pro Tools may take a little longer time to begin recording. To avoid this delay, you can put Pro Tools in "Prime for Record" mode first. Click Record in the Transport so the Record button flashes, and then Option-click (Mac) or Alt-click (Windows) on the Play button. The Stop button lights, and both the Play and Record buttons flash to indicate that "Prime for Record" mode is enabled. Now, when you click Play, Pro Tools will begin recording instantaneously.

Other Record Modes

Pro Tools has several other record modes that you can select using the Options menu: Destructive Record, Loop Record, QuickPunch, TrackPunch (Pro Tools HD only), and DestructivePunch (Pro Tools HD only). You can switch between these by Control-clicking (Mac) or Right-clicking (Windows and Mac) the Record Enable button in the Transport window. When you cycle through these modes, the Record Enable button changes to indicate the currently selected mode, adding a "D" to indicate Destructive, a loop symbol to indicate Loop Record, a "P" to indicate QuickPunch, a "T" to indicate TrackPunch, and "DP" to indicate DestructivePunch.

Destructive Record mode works like a conventional tape recorder where new recordings onto a particular track replace any earlier recordings to that track. There is little justification for using this mode unless you are running out of hard disk space to record to.

Loop Record mode lets you record multiple takes into the same track over a selected time range. Each successive take will appear in the Region List and can be placed in the track using the Takes List pop-up menu. This is very useful when a singer or musician is trying to perfect a difficult section.

QuickPunch lets you manually punch in and out of record on record-enabled audio tracks during playback by clicking the Record button in the Transport window. This is useful when the engineer or producer wants to decide "on-the-fly" which bit of a performance to replace while the singer or musician plays throughout the session.

TrackPunch mode lets you manually punch single tracks in and out, or take tracks out of record-enable, without interrupting online recording and playback. You can also simultaneously punch multiple tracks in and out.

DestructivePunch is a *destructive* recording mode that lets you manually punch tracks in and out during playback, replacing the audio material within any track that you punch into and thereby destroying what was there earlier. Like TrackPunch mode, DestructivePunch mode lets you punch tracks in and out individually, or punch multiple tracks in and out simultaneously, or take tracks out of record-enable, without interrupting online recording and playback.

Record Safe Mode

If you are worrying that you might accidentally put the wrong track or tracks into record and mess up an earlier recording, then be assured that it is very unlikely that this could happen with Pro Tools. Any new recordings are made to new files by default, so the earlier files will still be there in your Regions list and in your session's audio folder. This is what is referred to as "nondestructive" recording. Nevertheless, Pro Tools does have a "destructive" recording mode in which new recordings replace earlier recordings in the same files.

So, to prevent accidents, Pro Tools provides a Record Safe mode for each track that prevents tracks from being record-enabled. Simply Command-click (Macintosh) or Control-click (Windows) the track's Record Enable button, and this will become greyed out and won't let you enable the track to record. If you change your mind, just do this again to get out of Record Safe mode. If you hold the Option key (Macintosh) or Alt (Windows) at the same time, all tracks will be affected. And if you hold the Shift key as well, just the currently selected tracks will be affected.

Half-Speed Recording and Playback

A trick which recording engineers have been using for at least 50 years is to record a difficult-to-play musical part at half speed an octave below. When played back at normal speed, the part plays back at the correct tempo, but pitched up an octave—back to where it should be.

I remember being introduced to this technique in the early 1980's while recording to analog tape. I was struggling to play a tricky Clavinet part, and the recording engineer just ran the tape at half-speed and told me to play along an octave below. When he played, what I had recorded back at normal speed it sounded perfect—much tighter timing-wise than I had actually played it even at half-tempo. I subsequently learned that George Martin used this technique extensively when recording with The Beatles!

Now, you can use this technique when recording with Pro Tools. Press Command-Shift-Spacebar (Mac) or Control-Shift-Spacebar (Windows) when you start recording, and Pro Tools will play back existing tracks and record incoming audio at half-speed.

> ## Tip
>
> If you just want to play back a Pro Tools session at half-speed, all you need to do is press Shift-Spacebar. This can be very useful when playing along with or transcribing what is being played on existing recordings, which many musicians need to do from time-to-time.

Recording New Takes Using Playlists

Playlists provide a simple way to be able to keep recording new takes into the same track without any fuss. You can create a new playlist for the track each time you want to record a new take. This saves you the trouble of inserting or choosing a new track and then having to set the track up with the correct inputs and outputs, headphone mixes for the musicians, and inserts or plug-ins. Working on the same track you just recorded onto with a new playlist leaves the earlier take just as it was—but hidden from view and disabled from playing—so you can simply record another take into this new playlist. Every track lets you create as many of these edit playlists as you like, so if you want to record another take, you just create another new playlist.

Each track in the Edit window has a pop-up Playlist selector next to the track name. Click and hold this to reveal the Playlist menu. You can select "New…" to create a new Playlist. The "Duplicate" command is useful when you start editing your playlists later on, as is the "Delete Unused…." command. All the playlists you have created for this track are also listed here, and you can even access playlists for other tracks.

Figure 5.19 Pop-up Playlist selector.

When you select "New…," you are presented with a dialog box that lets you type a name for the new playlist or you can just OK this and go with the default name if this is fine for you.

Figure 5.20
Naming a new
Playlist.

Having created a new playlist, the first thing you will notice is that the track is now empty, and there are no regions visible, and the earlier take does not play back. So, you can go ahead and record-enable the track, ready to record the next take.

> **Tip**
>
> Playlists are also very useful for managing differently edited versions of a particular recording. You can chop up a recorded file one way in one playlist and another way in another playlist, and then swap these whenever you like while working on your arrangement.

Automatic Punch-in and Punch-out

Because Pro Tools records nondestructively by default, you don't have to take care to punch in and punch out exactly at the right places to avoid recording over material before and after the punch locations, unless you are using Destructive Record mode. All you need to do is to start recording anywhere before the section you want to replace and stop recording anywhere after the section you want to replace. Then, you simply trim the newly recorded and earlier recorded regions so that you replace the section of interest from the original recording.

Nevertheless, you can always drop in (punch in) on a track by specifying a range to record to first and setting up a pre-roll and post-roll (in the Transport window) as you would do with a conventional multitrack recorder. Playback will start at the pre-roll time, and Pro Tools will drop into record at the punch-in point and drop out of record at the punch-out point, stopping at the end of the post-roll time. Engineers who are more used to using multitrack tape recorders will probably feel more comfortable with this way of working. And if you really want to "re-live" the experience of working with tape recorders, you can enable Destructive Record to permanently replace the earlier recorded audio.

The simplest way to set up a range to record is to select a range in a track's playlist or in a Timebase ruler at the top of the Edit window using the Selection tool—making sure that the Edit and Timeline selections are linked. As usual, there are other ways to do this—such as typing the start and end times into the Transport window. Set pre- and post-roll times if you wish, then go ahead and record.

If you are recording in the default nondestructive mode, a new audio file is written to your hard drive and a new audio region appears in the record track and Region List. If you are recording in Destructive Record mode, the new audio overwrites the earlier material in the existing audio file and region.

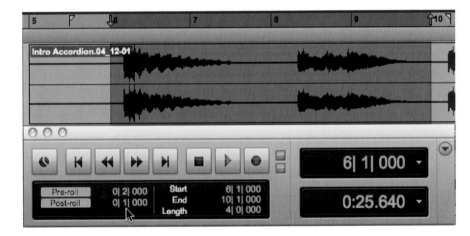

Figure 5.21
Automatic
Punch-in and
Punch-out with
Pre- and Post-roll
enabled.

Pre- and Post-Roll

Pre- and post-roll amounts can be entered in the Transport window or set from a track's playlist or Timebase ruler. Small yellow flags appear in the Main Time Scale ruler to indicate the amounts of pre- and post-roll that you have set. When pre- and post-roll are enabled, the flags are yellow, otherwise they are white.

Tip

You can always set Pre- and Post-Roll amounts, store them in a Memory Location, and recall the Memory Location whenever you like.

To set and enable the pre- and post-roll directly in a playlist, first make sure that "Link Timeline and Edit Selection" is enabled, and then with the Selector tool enabled, drag the cursor across the range that you want to select in the track's playlist.

> ### Note
>
> When you make an Edit selection with Timeline and Edit Selection linked, a pair of Timeline Selection Start and End markers will appear in the Timebase ruler. You can drag either of these to lengthen or shorten your selection. And if you Option-drag (Mac) or Alt-drag (Windows), you can move the selection backwards or forwards along the timeline.

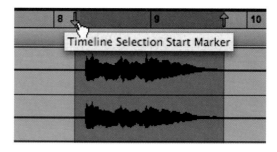

Figure 5.22
A linked Timeline and Edit selection.

With the Selector tool still enabled, Option-click (Mac) or Alt-click (Windows) in the track's playlist before the selection to enable the pre-roll and after the selection to enable the post-roll at those locations. Yellow flags will appear in the Timebase ruler.

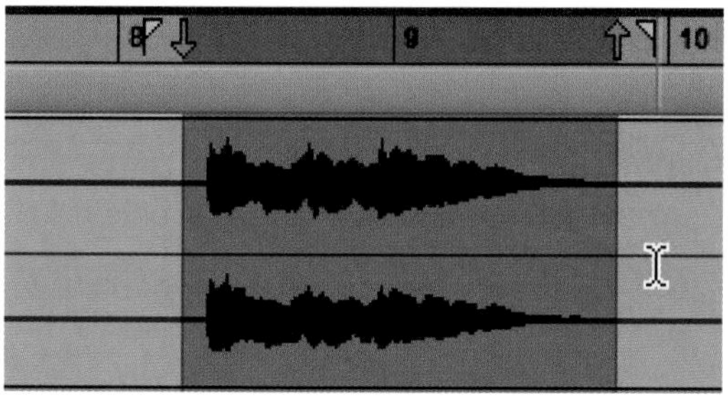

Figure 5.23
Pre- and Post-roll enabled with yellow flags showing to indicate these.

If you click on the pre- and post-roll buttons in the Transport window to deselect these, but leave the pre-and post-roll amounts set, the flags in the Timebase ruler will turn white to indicate this.

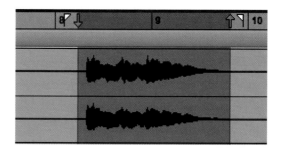

Figure 5.24
White flags indicate that Pre- and Post-roll amounts are set, but disabled.

To remove the pre- and post-roll amounts directly in a playlist, with the Selector tool enabled, you can simply Option-click (Mac) or Alt-click (Windows) near the start of the Edit selection to disable the pre-roll and near the end of the Edit selection to disable the post-roll.

Another way to remove the pre- and post-roll amounts is to drag the pre- and post-roll flags until they coincide exactly with the Timeline Selection Start and End markers.

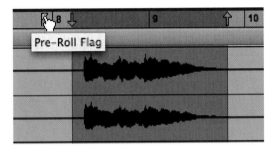

Figure 5.25
Drag the Pre-or Post-roll flags along the Timeline to set or remove the amounts.

The Pre- and Post-roll Flags can be moved whenever you like by dragging these in the Main Timebase ruler, either separately or at the same time, and if you set the Edit mode to Grid, the Pre- and Post-roll flags will snap to the current Grid value when you drag them along the Timeline.

To set pre- and post-roll values to the same amount, Option-drag (Mac) or Alt-drag (Windows) either the pre- or the post-roll flag in the ruler. The other flag will immediately be set to the same value and will adjust to the identical amount as you drag the selected flag.

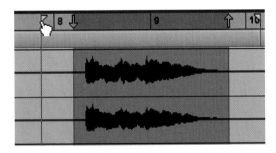

Figure 5.26
Option-drag the Pre- or Post-roll flags to set identical amounts.

Destructive Record

If you prefer to have Pro Tools behave more like a conventional multitrack recorder, where you typically record additional takes over earlier takes, you can always enable the Destructive Record mode from the Operations menu. The letter "D" will appear in the Record button on the Transport to remind you that you are using this mode. When you have recorded the first take, simply leave the track record-enabled, go back and record again. Your new recording will have replaced whatever you had earlier recorded into that audio file on disk.

> ### Note
>
> If you insert the cursor at the end of your earlier recording, the additional material will be appended at the end of the file—thus extending the length of the track.

Now, that we have all got used to the idea of Pro Tools hard disk recording being nondestructive (as long as you always record new takes to new files on disk, which is the default situation), it may seem a little strange for some of you to use the Destructive Record mode. Experienced recording engineers used to work with conventional multitrack recorders will, of course, be perfectly familiar with this way of working. Don't forget, this can be a more efficient way to work – especially if you know what you are doing and are confident that you are not too likely to make a mistake. You won't have to take the time and trouble to sort through the alternate takes and delete them from your hard drive afterwards, and if you are running low on hard disk space Destructive Record can be a boon.

Loop Recording Audio

If you are trying to pin down that perfect 8-bar instrumental solo in the middle of your song or the definitive lead vocal on the verse, it's time to use the Loop Recording feature. To set this up is very straightforward. Just select Loop Record from the Operations menu, and you will see a loop symbol that appears on the Record button in the Transport window. Make sure you have Link Edit and Timeline Selection checked in the Operations menu, and then use the Selector tool to drag over the region you want to work with in the Edit window. You can set a pre-roll time if you like, or you can simply select a little extra at the beginning and then trim this back later. Now, when you hit record, Pro Tools will loop around this selection, recording each take as an individual region within one long file. When you have finished recording you can choose the best take at your leisure. All the takes will be placed into the Audio Regions list and numbered sequentially, with the last one left in the track for you. Now, if you want to hear any of the other takes, just select the last take with the Grabber tool and Command-drag (Macintosh) or Control-drag (Windows) whichever take you fancy from the Audio Regions list, and it will automatically replace the selected take in the track—very convenient! An even faster way is to Command-click (Mac) or Control-click (Windows) with the Selector tool at the exact start of the loop or punch range. This immediately brings up the Takes List pop-up—making it even easier to select alternate takes.

But what if you want to audition takes from an earlier session? These will not be listed here normally, as the start times are likely to be different. The User Time Stamp for each take in Loop record is set to the same start time, at the beginning of the loop, and the Takes List is based on matching start times. That is why it displays all your takes when you bring the List up at the start time of your loop. So, if you want to include other takes from an earlier session in the Takes List pop-up for a particular location, you can simply set the User Time Stamp for these regions to the same as for these new takes, and they will all appear in the Takes List. And if you plan on recording some more takes later on for this same section, you should store your loop record selection as a Memory Location. This way, these takes will also appear in the Takes List pop-up for that location.

You can also restrict what appears in the Takes List according to the Editing Preferences you choose. If you enable "Take Region Name(s) That Match Track Names" then the list will only include regions that take their name from track/playlist. This can be useful when sorting through many different takes from other sessions, for example. Or maybe you want to restrict the Takes List to regions with exactly the same length as the current selection. In this case, make sure that you have enabled "Take Region Lengths That Match" in the Editing Preferences.

If you have both of these preferences selected you can even work with multiple tracks to replace all takes on these simultaneously. When you choose a region from the Takes List in one of the tracks, not only will the selected region be replaced in that track but also the same take numbers will be placed in the other selected tracks.

A third option is provided in the Editing Preferences to make any "Separate Region" commands you apply to a particular region simultaneously to all other related takes, that is, takes with the same User Time Stamp. You could use this to separate out a particular phrase that you want to compare with different takes, for example.

> ## Note
> All the regions in your session with the same User Time Stamp will be affected unless you keep one or both of the other two options selected, in which case, the Separate Region command will only apply to regions that also match these criteria.

> ## Tip
> It can be easy to forget that you have left this preference selected and end up accidentally separating regions when you don't intend to, so make sure you deselect this each time after using it.

Using Loop Recording

First, make sure that Loop Record is selected from the Options menu so that the loop symbol is showing in the Transport window's Record Enable button. If you are going to set a record or play range by selecting within a playlist, make sure that the Edit and Timeline selections are linked. (Select Link Timeline and Edit Selection from the Options menu or make sure that the Link Timeline and Edit Selection button above the rulers is highlighted.) Choose the Selector Tool and drag the cursor across the range of audio in the playlist that you want to loop record over. Alternatively, you can set Start and End times for the loop in the Transport window. The Loop Record selection must be at least 1 second long.

> **Tip**
>
> Although you can set a pre-roll time that will be used on the first pass, and a post-roll time that will be used on the last pass, I recommend that you simply select a loop range that includes some time before and some time after the range you wish to record over. Later, you can trim back the recorded takes to the proper length with the Trim tool.

Record-enable the track or tracks you want to record onto, click Record and Play in the Transport window when you are ready to begin recording, and click Stop in the Transport window when you have finished.

Figure 5.27
Setup for Loop Recording.

All the takes are recorded into a single audio file with sequentially numbered regions defined for each take. The most recently recorded take is left in the record track. All of these takes appear as regions in the Region List, and you can audition takes on their own, from the Region List, or from the Takes List pop-up menu.

Selecting Takes after Loop Recording

Each take created during a Loop Record session is given the same User Time Stamp. So, to select a different take from the Region List, you can use the Time Grabber tool to select the current take in the Edit window, and then Command-drag (Macintosh) or Control-drag (Windows) another take from the Region List into the playlist. This will immediately replace the earlier take, snapping exactly to the correct location because it has the same User Time Stamp.

To access the Takes List pop-up menu to select a different take, make sure that the Selector tool is engaged, select the take currently residing in the track, and then Command-click (Mac) or Control-click (Windows) anywhere in the region. The pop-up menu that appears contains a list of regions that share the same User Time Stamp. So, when you choose a region from the Takes List pop-up menu, this replaces the earlier take and snaps exactly to the correct location in the same way.

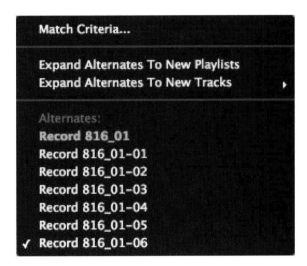

Figure 5.28
The Takes List
pop-up menu.

> **Tip**
>
> If you intend to use the track compositing feature after your Loop Recording session, you should enable the "Automatically Create New Playlists when Loop Recording" preference which you will find in the Record section of the Operation Preferences window. This copies alternate takes created during loop recording to new playlists in the track.

Manual Punch-in and Punch-out

Pro Tools provides three different manual punch recording modes that all have their uses when you are tracking and overdubbing music.

QuickPunch is a nondestructivepunch-recording mode that lets record-enabled tracks be punched in and out during playback by clicking the Record button in the Transport and provides instantaneous monitor switching on punch-out.

TrackPunch (Pro Tools HD only) is a nondestructivepunch-recording mode that lets individual tracks be punched in, punched out, and taken out of record-enable without interrupting online recording and playback. TrackPunch also provides instantaneous monitor switching on punch-out.

DestructivePunch (Pro Tools HD only) is a destructivepunch-recording mode that permanently replaces pre-existing track material with newly recorded audio in the original audio file. Pro Tools only records while tracks are punched in, inserting new material and replacing audio in the track playlist. As with TrackPunch mode, individual tracks can be punched in, punched out, and taken out of record enable without interrupting online recording and playback.

With both TrackPunch and DestructivePunch, you can record-enable tracks on-the-fly; punch tracks in and out using on-screen record-enable buttons, or remotely from a synchronizer, or from a control surface, or using a foot switch; and compare and match levels using TrackInput switching.

Note

QuickPunch and TrackPunch create new audio files for each *pass* (a pass is one cycle of starting and stopping the Transport). Pro Tools begins recording a new audio file when playback begins, automatically generating *regions* in that file at each punch-in and punch-out point. Up to 200 "running punches" can be performed in a single pass. After recording, the new audio regions appear in the Region List. This includes the whole-file audio region encompassing all punches from the record pass, along with the regions derived for each punch.

Voice Requirements for Punch Recording

With Pro Tools|HD systems, the punch recording modes (QuickPunch, TrackPunch, and DestructivePunch) use an additional voice for each record-enabled mono track, so the total number of available voices for punch recording is cut in half. On a large project, it would not be impossible to find yourself short of voices, but you will be prompted to free up the necessary voices if this happens. Don't forget that you can always turn off the voice assignments to less-important tracks—or make these inactive. Voices that are being used by other tracks, but are not record-enabled, will be claimed for use during punch recording as necessary, so you will not hear these tracks play back.

With TDM systems, the recommended procedure is to use Dynamically Allocated voice assignments for each track so that Pro Tools will automatically distribute the voices between the DSP chips. For example, for a 128-voice configured Pro Tools|HD system, Dynamically Allocated voicing distributes voices evenly across four sets of voices (1–32, 33–64, 65–96, and 96–128). Similarly, with a 192-voice configured Pro Tools|HD Accel system, Dynamically Allocated Voicing distributes voices evenly across four sets of voices (1–48, 49–96, 97–144, and 145–192).

However, it can still be a good idea to assign voices to tracks manually, especially if you anticipate running out of voices and want to make sure that the most important tracks will always play.

> **Note**
>
> If you decide not to use Dynamically Allocated Voicing, the track voices must be set manually to distribute the tracks across the appropriate number of DSP chips. For example, to use QuickPunch on 32 tracks without Dynamically Allocated Voicing, tracks 1–16 must be assigned to voices 1–16 and tracks 17–32 must be assigned to voices 33–48.

Crossfades while Punch Recording

Pro Tools automatically writes a crossfade for each punch point when using QuickPunch and TrackPunch to make sure that there are smooth transitions between punch regions. You can set the length for these crossfades using the QuickPunch/TrackPunch Crossfade Length option that you will find on the Editing Preferences page. If this is set to zero, Pro Tools won't create any crossfades at the punch-in/out points. If you set a nonzero value, Pro Tools writes a pre-crossfade (up to but not into the punched region boundary) at the punch-in point, and a post-crossfade at punch-out (after the punched region). A typical value to use for the crossfade length for punches would be 10 milliseconds. If necessary, you can always edit the crossfades when you have finished punch recording, using the standard fade editing features.

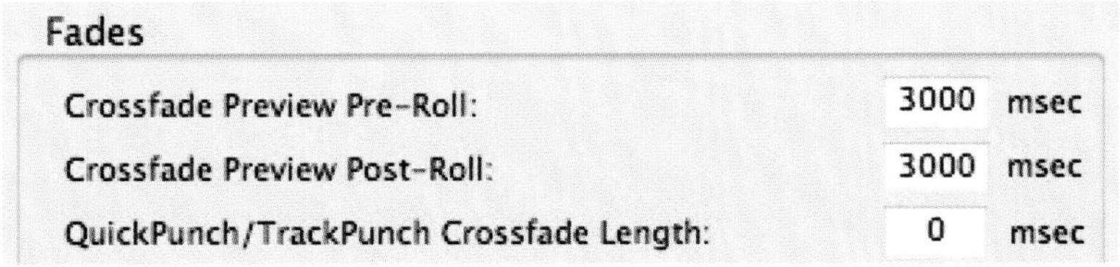

Figure 5.29
Fades Editing
Preferences.

Fades		
Crossfade Preview Pre-Roll:	3000	msec
Crossfade Preview Post-Roll:	3000	msec
QuickPunch/TrackPunch Crossfade Length:	0	msec

> **Note**
>
> Regardless of the current QuickPunch/TrackPunch Crossfade Length setting, Pro Tools always executes a 4 millisecond "monitor only" crossfade (which is not written to disk) to avoid distracting pops or clicks that might occur as you enter and exit record mode.

DestructivePunch doesn't give you a choice of crossfade lengths: a fixed 10-millisecond linear crossfade is automatically created at each in and out point.

Setting Levels for Punch Recording

Before you start punch recording, you may need to adjust the monitoring levels for the incoming audio and the playback audio for the tracks you will punch record onto so that the level that you hear in the control room monitors (and the level that the performer hears in the headphones) does not change drastically when you punch in.

To get the playback level for a track from disk to match the monitoring level of incoming audio, start playback and compare the levels of the input source with the audio on disk, clicking the TrackInput button to toggle the track source between input and playback from disk. The TrackInput button turns green when it is monitoring the input. When you have set all the levels appropriately, you are ready to start punch recording.

QuickPunch

If you like to work fast while doing overdubs, you can use the QuickPunch feature to drop in and out of record on record-enabled tracks. Normally, when you use Auto Input Monitoring then switch back to monitoring track material on punch-out, this is not instantaneous. The big difference here is that QuickPunch instantly switches from monitoring the input to monitoring the playback when you punch-out of recording. Also, with QuickPunch, Pro Tools actually starts recording a new audio file as soon as you start playback, automatically defining and naming regions in that file at each punch-in/out point.

> ### Tip
>
> If a musician plays or a vocalist sings something that you would like to keep while the Transport is running in QuickPunch mode, but before you have punched in, all you need to do is punch in later than this, and when you stop recording, trim the region backwards along the Timeline to reveal the earlier material, which will have been recorded!

Preparing for QuickPunch

You can enable QuickPunch mode by selecting this in the Options menu or by Right-clicking the Record button in the Transport window and selecting QuickPunch from the pop-up menu. If you prefer a keyboard command, you can use Command-Shift-P (Mac) or Control-Shift-P (Windows). Possibly the fastest way is to Control-click (Mac) or Start-click (Windows) on the Record button in the Transport window to cycle through available Record modes until QuickPunch mode is selected. When QuickPunch mode is enabled, a "P" appears in the Record button in the Transport.

Using QuickPunch

To record using QuickPunch mode, first make sure that the individual tracks that you want to record onto are also record-enabled, with their Record-enable buttons flashing red. Cue Pro Tools to the location you want to start from, using a pre-roll value if you like, and then press the Spacebar or click the Play button to start playback. When you reach the location at which you want to start recording, simply click the Record button in the Transport window. Pro Tools will immediately switch to monitoring the input (if it is not already in Input Monitoring mode) and will start recording. While a track is recording, its Record Enable button lights solid red. To punch out again, just click the Record button in the Transport window. You can continue to punch in and out up to 200 times during a single pass. When you have finished recording, just click Stop in the Transport window or press the Spacebar to stop recording and playback.

Figure 5.30
Pro Tools in QuickPunch mode with a track Record-enabled—good to go!

TrackPunch (Pro Tools HD Only)

TrackPunch mode lets you punch tracks in and out individually, or punch multiple tracks in and out simultaneously, without interrupting online recording and playback. Up to 200 "running punches" can be performed in a track during a single TrackPunch pass.

Preparing for TrackPunch

You can enable TrackPunch mode by selecting this in the Options menu or by Right-clicking the Record button in the Transport window and selecting TrackPunch from the pop-up menu. If you prefer a keyboard command, you can use Command-Shift-T (Mac) or Control-Shift-T (Windows). Possibly the fastest way is to Control-click (Mac) or Start-click (Windows) on the Record button in the Transport window to cycle through available Record modes until TrackPunch mode is selected. When TrackPunch mode is enabled, a "T" appears in the Record button in the Transport.

Before each pass, you must TrackPunch-enable all the tracks that you intend to punch. You can enable tracks for TrackPunch without record enabling them. This lets you punch in on individual tracks at any time after starting playback by clicking their respective Record Enable buttons. Or, you can simultaneously TrackPunch-enable tracks and record-enable them. This lets recording begin on all the enabled tracks as soon as the transport is record-armed and playback begins.

To TrackPunch-enable a single audio track: Control-click (Mac) or Start-click (Windows) on the track's Record Enable button to toggle the button to solid blue. To TrackPunch-enable or disable *all* audio tracks: Option-Control-click (Mac) or Alt-Start-click (Windows) on any track's Record Enable button to toggle all the Record Enable buttons to solid blue. To TrackPunch-enable or disable all *selected* audio tracks: Option-Control-Shift-click (Mac) or Alt-Start-Shift-click (Windows) on any selected track's Record Enable button to toggle the Record Enable buttons for the selected audio tracks to solid blue. These actions TrackPunch-enable the tracks, but do not record-enable the tracks, so you will still need to click on the individual tracks that you wish to punch in on.

Or, you can simultaneously TrackPunch-enable tracks *and* record-enable them. This lets recording begin on all the enabled tracks as soon as the transport is record-armed and playback begins.

To simultaneously TrackPunch-enable and record-enable a single audio track: Click the track's Record Enable button so that it flashes from blue to red. To simultaneously TrackPunch-enable and record-enable all audio tracks: Option-click (Mac) or Alt-click (Windows) any track's Record Enable button so that all tracks' Record Enable buttons flash from blue to red. To simultaneously TrackPunch-enable and record-enable all selected audio tracks: Option-Shift-click (Mac) or Alt-Shift-click (Windows) any selected track's Record Enable button so that the Record Enable buttons for the selected audio tracks flash from blue to red.

After putting the Transport Record button into TrackPunch mode, when you first click on any individual track's record enable button, this will both TrackPunch-enable it and record-enable the track. To confirm this, the small Record Enable Status LED to the right of the Transport Record button goes red. Also, when a track is both TrackPunch-enabled and record-enabled, the track's record-enable button will flash from blue to red.

Figure 5.31 When a track is TrackPunch-enabled and is also record-enabled, its Record Enable button lights flashes from blue to red, the Transport window Record button also flashes blue to red and has a "T" in the middle, and the small Record LED to the right of this *is* lit.

If you click once more on the track's record-enable button, this will disable it for recording, but will leave it TrackPunch-enabled, and the Record Enable Status LED in the Transport window will be unlit. With TrackPunch enabled (but not record-enabled) for a track, the Track's record-enable button will light up solid blue.

Figure 5.32 When a track is TrackPunch-enabled, but not record-enabled, its Record Enable button lights solid blue, the Transport window Record button flashes blue to red and has a "T" in the middle, and the small Record LED to the right of this *is not* lit.

To disable a track that you have TrackPunch-enabled, so that it is neither TrackPunch-enabled nor record-enabled, Control-click (Mac) or Start-click (Windows) on the track's record-enable button. This will no longer be lit and, when no tracks are TrackPunch- or record-enabled, neither will the Record button in the Transport window.

Note

If at least one track is TrackPunch-enabled, the Record button in the Transport window lights solid blue. When TrackPunch mode is enabled but no tracks are TrackPunch-enabled, the Transport Record button flashes from grey to red. If at least one track is TrackPunch-enabled, the Transport Record button flashes from blue to red. If at least one TrackPunch-enabled track is also record-enabled, the Transport Record button flashes from blue to red, *and* the small record LED to the right of this also lights up red. Whenever at least one audio track is recording, the Transport Record button lights up solid red.

Using TrackPunch

When you are all set up to use TrackPunch, you can either punch in on individual tracks or punch in on multiple tracks simultaneously when you reach the first punch-in point during playback. You may also want to punch in immediately, then punch out further into the track, then punch in again later on.

To punch in on individual tracks, make sure that each track you want to punch in on is TrackPunch-enabled, with its Record Enable button lit solid blue. Click Record in the Transport to enter the TrackPunch Record Ready mode, with the Record button flashing from blue to red. Then, click Play in the Transport to begin playback. While the session plays back, you can punch in and out on individual TrackPunch-enabled tracks by clicking their record-enable buttons.

Tip

After a TrackPunch recording pass, the punched track's playlist in the Edit window displays the regions created by punching. You can use any of the Trim tools after punch recording to open up the head or tail of TrackPunch recorded regions, or to reveal the parent audio file that was recorded in the background. This lets you compensate for any late or missed punches.

To punch in on multiple tracks simultaneously, make sure that each track you want to punch in on is both TrackPunch-enabled and Record-enabled. To punch in or out on all the TrackPunch-enabled tracks simultaneously, play back the session and click the Record button in the Transport window whenever you like.

> **Note**
>
> Alternatively, click Record in the Transport first, and then Option-Shift-click (Mac) or Alt-Shift-click (Windows) on any selected track's Record Enable button to simultaneously punch in and out on all the currently selected TrackPunch-enabled tracks.

To start your TrackPunch pass in record, punching in immediately on all tracks: make sure that all the tracks are both TrackPunch-enabled and record-enabled and that the Transport is in TrackPunch Record Ready mode, and then click Play in the Transport to begin playback. During playback, you can punch out and then punch in and out again whenever you like on any TrackPunch-enabled track by clicking its Record Enable button.

DestructivePunch (Pro Tools HD Only)

DestructivePunch is a destructive version of TrackPunch mode. Where TrackPunch always records audio to a new file, DestructivePunch destructively records audio directly into the original file, using a fixed 10-millisecond linear crossfade. No additional regions are created when recording in DestructivePunch mode. Up to 200 "running punches" can be performed in a track during a single DestructivePunch pass.

Preparing for DestructivePunch

To use DestructivePunch on an audio track, the track must contain a continuous audio file of a minimum length, which you can set in the Pro Tools Operation preferences page. Also, the file must start at the beginning of the session (at sample 0) and the File Length must be equal to or greater than the DestructivePunch File Length setting. If the DestructivePunch File Length is not sufficiently long for the files you are working with, it is easy enough to change this in the Operation Preferences so that it is equal to or greater than the length of the current file.

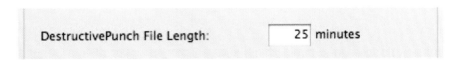

DestructivePunch File Length: 25 minutes

Figure 5.33 Setting the DestructivePunch File Length in the Operation Preferences.

Sometimes you may want to use DestructivePunch with a file that does not start at the beginning of the session and that you cannot move to the start of the session because this would put it out of sync with other files. And very often a playlist will contain regions taken from different files. In these cases you can either use the Bounce to Disk command or use the Consolidate command to create a continuous file of the required length that starts at the beginning of the session. Because this is such a common situation, Pro Tools has a special "Prepare DPE Tracks" command in the Options menu that lets you consolidate audio on all DestructivePunch-enabled tracks. To use this command, you need to enable DestructivePunch mode first, then make sure that all the tracks you want to prepare are DestructivePunch-enabled, but not record-enabled. When you use the "Prepare DPE Tracks" command, Pro Tools consolidates audio on all DestructivePunch-enabled tracks from the beginning of the session to the value specified in the DestructivePunch File Length preference.

You can enable DestructivePunch mode by selecting this in the Options menu or by Right-clicking the Record button in the Transport window and selecting DestructivePunch from the pop-up menu. A faster way is to Control-click (Mac) or Start-click (Windows) on the Record button in the Transport window to cycle through available Record modes until DestructivePunch mode is selected. When DestructivePunch mode is enabled, "DP" appears in the Record button in the Transport.

Before each pass, you must DestructivePunch-enable all tracks that you intend to punch. You can enable tracks for DestructivePunch without record enabling them. This lets you punch in on individual tracks at any time after starting play-back by clicking their respective Record Enable buttons. You may also need to do this if you want to use the Prepare DPE Tracks command to consolidate files for DestructivePunch recording.

To DestructivePunch-enable a single audio track: Control-click (Mac) or Start-click (Windows) the track's Record Enable button to toggle the button to solid blue. To DestructivePunch-enable or disable *all* audio tracks: Option-Control-click (Mac) or Alt-Start-click (Windows) any track's Record Enable button to toggle all Record Enable buttons to solid blue. To DestructivePunch-enable or disable all *selected* audio tracks: Option-Control-Shift-click (Mac) or Alt-Start-Shift-click (Windows) a selected track's Record Enable button to toggle the Record Enable buttons for the selected audio tracks solid blue. These actions DestructivePunch-enable the tracks, but do not record-enable the tracks, so you will still need to click on the individual tracks that you wish to punch in on.

Or, you can simultaneously DestructivePunch-enable tracks *and* record-enable them. This lets recording begin on all the enabled tracks as soon as the transport is record-armed and playback begins.

To simultaneously DestructivePunch-enable and record-enable a single audio track: Click the track's Record Enable button so that it flashes from blue to red. To simultaneously DestructivePunch-enable and record-enable all audio tracks: Option-click (Mac) or Alt-click (Windows) a track's Record Enable button so that all tracks' Record Enable buttons flash from blue to red. To simultaneously DestructivePunch-enable and record-enable all selected audio tracks: Option-Shift-click (Mac) or Alt-Shift-click (Windows) any selected track's Record Enable button so that the Record Enable buttons for the selected audio tracks flash from blue to red.

After putting the Transport Record button into DestructivePunch mode, when you first click on any individual track's Record Enable button, this will DestructivePunch-enable it and record-enable the track. To confirm this, the small Record Enable Status LED to the right of the Transport Record button goes red. Also, when a track is both DestructivePunch-enabled and record-enabled, the track's Record Enable button will flash from blue to red.

Figure 5.34 Record Enable Status LED.

If you click once more on the track's record-enable button, this will disable it for recording but will leave it DestructivePunch-enabled, and the Record Enable Status LED in the Transport window will be unlit. With DestructivePunch enabled (but not record-enabled) for a track, the Track's Record Enable button will light up solid blue.

Figure 5.35 DestructivePunch Is enabled in the Options menu and the Record button has been selected in the Transport; an Audio track is also Record-enabled: so the Transport Record button flashes blue to red, the individual track record-enable buttons flash blue to red, and the small LED to the right of the Record button in the Transport window lights up red.

To disable a track that you have DestructivePunch-enabled so that it is neither DestructivePunch-enabled nor record-enabled, Control-click (Mac) or Start-click (Windows) on the track's record-enable button. This will no longer be lit, and when no tracks are DestructivePunch- or record-enabled, neither will the Record button in the Transport window.

> **Note**
>
> If at least one track is DestructivePunch-enabled, the Record button in the Transport window lights solid blue. When DestructivePunch mode is enabled but no tracks are DestructivePunch-enabled, the Transport Record button flashes from grey to red. If at least one track is DestructivePunch-enabled, the Transport Record button flashes from blue to red. If at least one DestructivePunch-enabled track is also record-enabled, the Transport Record button flashes from blue to red, *and* the small record LED to the right of this also lights up red. Whenever at least one audio track is recording, the Transport Record button lights up solid red.

Using DestructivePunch

When you are all set up to use DestructivePunch, you can either punch in on individual tracks or punch in on multiple tracks simultaneously when you reach the first punch-in point during playback. You may also want to punch in immediately, and then punch out further into the track, and punch in again later on.

To punch in on individual tracks, make sure that each track you want to punch in on is DestructivePunch-enabled, with its Record Enable button lit solid blue. Click Record in the Transport to enter the DestructivePunch Record Ready mode, with the Record button flashing from blue to red. Then, click Play in the Transport to begin playback. While the session plays back, you can punch in and out on individual DestructivePunch-enabled tracks by clicking their record-enable buttons.

To punch in on multiple tracks simultaneously, make sure that each track you want to punch in on is both DestructivePunch-enabled and record-enabled. To punch in or out on all the DestructivePunch-enabled tracks simultaneously, play back the session and click the Record button in the Transport window whenever you like.

> **Note**
>
> Alternatively, click Record in the Transport first, and then Option-Shift-click (Mac) or Alt-Shift-click (Windows) on any selected track's Record Enable button to simultaneously punch in and out on all the currently selected DestructivePunch-enabled tracks.

To start your DestructivePunch pass in record, punching-in immediately on all tracks: make sure that all the tracks are both DestructivePunch-enabled and record-enabled and that the Transport is in DestructivePunch Record Ready mode, and then click Play in the Transport to begin playback. During playback, you can punch out and then punch in and out again whenever you like on any DestructivePunch-enabled track by clicking its Record Enable button.

Delay Compensation and DestructivePunch

It is important to be aware of whether or not Delay Compensation was in use when the recording that you wish to use DestructivePunch with was originally recorded. If Delay Compensation was used on the original recording and this is not used when you punch in using DestructivePunch, or vice versa, the new audio

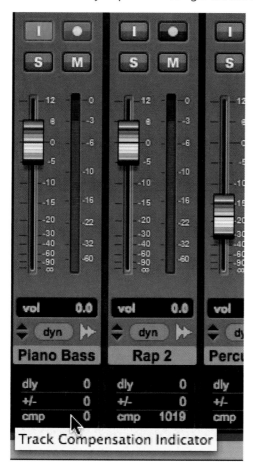

will not be recorded into exactly the correct position in the audio file. It will be offset in one direction or the other by the amount of the Delay Compensation. To make sure that this does not happen, if Delay Compensation was inactive when recording the original file, it should be deactivated while using DestructivePunch and if was active when recording the original file, it should be kept active while using DestructivePunch.

Consequently, you may need to manually override the automatic suspension of Delay Compensation that occurs when record-enabled tracks switch to Input Monitoring when you are using DestructivePunch.

This feature is called Auto Low Latency and can be turned off by Right-clicking the Track Compensation indicator on the track and selecting Auto Low Latency Off or by Command-Control-clicking (Mac) or Control-Start-clicking (Windows) on the Track Compensation indicator on the track.

Figure 5.36 Track Compensation is automatically suspended by the Auto Low Latency feature when Input Monitoring is enabled.

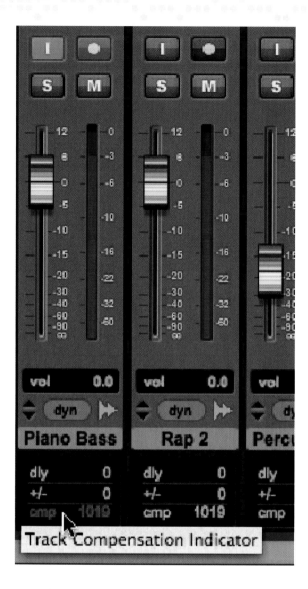

Figure 5.37 Track Compensation suspension can be overridden with a Command-Control-click (Mac) or Control-Start-click (Windows) on the Track Compensation indicator.

Tip

To apply Delay Compensation to all selected tracks where Delay Compensation was suspended, either Shift-Right-click the Track Compensation indicator on any selected track and select Auto Low Latency Off, or Command-Control-Shift-click (Mac) or Control-Start-Shift-click (Windows) on the Track Compensation indicator on any selected track.

After Punch Recording

When you are finished with the record pass after Punch recording, the track Record Enable status and transport Record Arm status for the tracks you have been recording onto will follow the current Audio Track RecordLock and Transport RecordLock settings. You can set these in the Record section of the Operation Preferences.

Figure 5.38
RecordLock
Preferences.

If the Audio Track RecordLock option is enabled, the record-enabled audio tracks remain armed when playback or recording stops. When the Audio Track RecordLock option is not enabled, record-enabled audio tracks are taken out of record enable when Pro Tools is stopped, as with a digital dubber.

If the Transport RecordLock preference setting is not enabled, the Transport Record disarms when Pro Tools is stopped. When enabled, the Transport Record remains armed when playback or recording stops, so you don't have to rearm the Transport between takes, as with a digital dubber.

So, for music production, you will normally have the Audio Track RecordLock preference selected to enable it and the Transport RecordLock preference unticked to disable it. If you are using Pro Tools as a digital dubber, then you would reverse these preferences.

Summary

If you have followed this chapter thoroughly, you will be able to set up Pro Tools to record in a variety of scenarios. Setting up headphone cue mixes; strategies for combining TDM and RTAS plug-ins if you are using a Pro Tools HD system; input monitoring; adjusting monitoring levels during record and playback; recording with or without effects; recording through plug-ins—you need to get all of this stuff sorted out before you get started with serious recording projects.

Whether you are overdubbing one musician at a time or recording a whole band, you should know how to arm tracks in the various record modes, how to record or play back at half-speed, how to punch in and out automatically or manually, how to loop record audio, and how to use playlists to record alternate takes.

You also need to understand how and where latency delays occur and develop strategies to avoid these delays that take into account the needs of the musicians you are recording, and the results that you are aiming to achieve. Get to know how the delay compensation features work and you will be on your way to mastering the Pro Tools recording environment. And to make those quick fixes that are needed with even the most practiced studio musicians, you need to master the various punch modes—choosing just the right length of pre-roll in bars and beats that gives the musicians enough time to prepare for the punch-in, without wasting time by starting too many bars before the punch location.

Pro Tools has all the features you need to record just about anything that makes a sound that can be captured by a microphone, but if you are new to this software environment, and even if you are upgrading from older systems, there is plenty that is new to learn here. It may take a while before you get an opportunity to use all these recording features, so feel free to revisit this chapter whenever you would like to remind yourself of how everything works.

In This Chapter

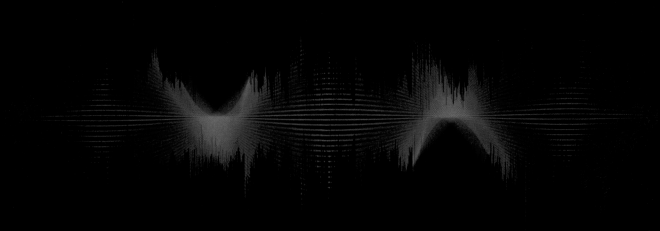

Editing

Introduction

To become proficient at editing using Pro Tools, you need a firm understanding of how the various Edit modes work and how to select material to edit using the various editing commands. Understanding what regions are and how to manipulate these is also crucially important. Knowing the best ways to create loops and edit these is particularly important when you are working in the various pop, rap, and dance music genres, where the Dynamic Transport mode can also be very helpful. For example, Track Compositing offers an easy-to-use workflow, but seasoned Pro Tools users may choose to continue using their previously tried and tested methods. As usual, there are several ways of achieving the same results with Pro Tools—it is one of the most flexible environments available for editing audio!

The Edit Menu Commands

You can access most of the commands that you will use for editing from the Edit menu, but you will also find some commands, such as Capture and Loop, that could be considered as editing commands in the Region menu and vice versa (I'm thinking about Trim Region and Separate Region here). Gaining familiarity with the Edit menu and the editing commands in the Region menu should be priorities for aspiring Pro Tools editors.

Some of these commands, such as Select All, don't require any additional explanation or are explained perfectly well in the Pro Tools Reference Guide, while explanations of others, such as the Automation editing commands belong elsewhere. One of the fastest ways to learn the keyboard commands is to look through these menus, note the keyboard equivalents for the various commands, then use these as often as possible from memory.

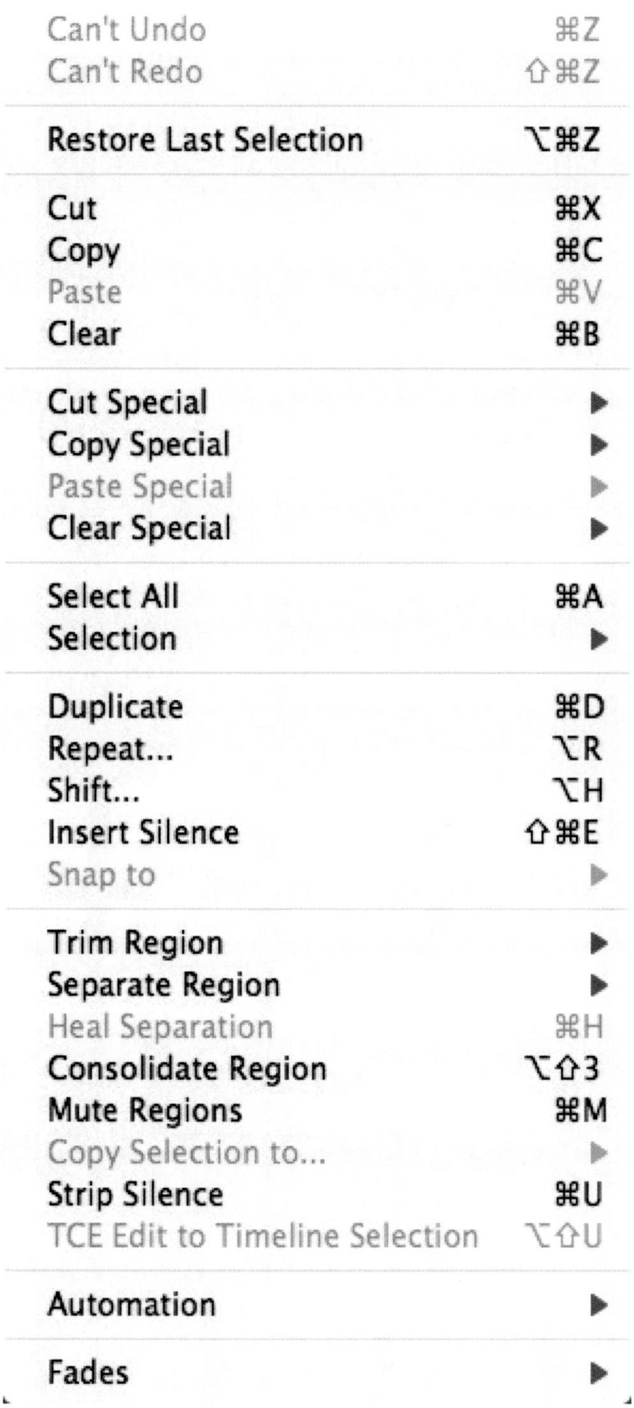

Can't Undo	⌘Z
Can't Redo	⇧⌘Z
Restore Last Selection	⌥⌘Z
Cut	⌘X
Copy	⌘C
Paste	⌘V
Clear	⌘B
Cut Special	▶
Copy Special	▶
Paste Special	▶
Clear Special	▶
Select All	⌘A
Selection	▶
Duplicate	⌘D
Repeat...	⌥R
Shift...	⌥H
Insert Silence	⇧⌘E
Snap to	▶
Trim Region	▶
Separate Region	▶
Heal Separation	⌘H
Consolidate Region	⌥⇧3
Mute Regions	⌘M
Copy Selection to...	▶
Strip Silence	⌘U
TCE Edit to Timeline Selection	⌥⇧U
Automation	▶
Fades	▶

Figure 6.1 Pro Tools HD Edit Menu.

Undo and Undo History

Pro Tools tracks the last 32 undoable operations that you carry out, storing these in a queue that you can step back through using the Edit menu's Undo command or by pressing Command-Z (Mac) or Control-Z (Windows). If you change your mind about this, you can step forward through the undo queue by choosing the Redo command from the Edit menu or by holding the Shift key while you press Command-Z (Mac) or Control-Z (Windows).

When you reach the 32-step limit and perform another undoable operation, the oldest stored operation is removed from the queue. Keep in mind that it is possible to set a lower limit for the number of undoable steps in the Editing Preferences, so if you are working on someone else's system, or if someone else may have reset this, you should check to make sure what limit is set.

You can also use the Undo History window that you will find in the Window menu. This lets you view the queue of undoable and redoable operations. It shows undoable operations in bold type, and if you click on any of these, it will undo all the undoable operations back to and including the one you clicked on, and will show these as redoable operations in italic type. Click on any redoable operation to redo all operations up to and including the one you clicked on.

The Undo History can optionally show edit creation times, which can be a help when you are deciding which operation to undo or redo. This option can be accessed using the pop-up selector at the top right corner of the Undo History window. You can also undo or redo all operations in the queue, or clear the whole queue.

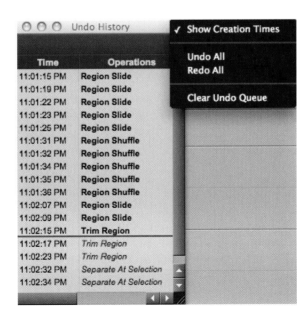

Figure 6.2
Undo History
window.

Restore Last Selection

Very often you will make an Edit or Timeline selection then lose it. Perhaps, you accidentally click somewhere in the Edit window without realizing, or whatever. Pro Tools has a really useful command to get you out of trouble here! Just choose "Restore Last Selection" from the Edit menu or press Command-Option-Z (Mac) or Control-Alt-Z (Windows) and your selection will magically reappear!

Cut, Copy, Paste, and Clear Commands

Pro Tools has some similarities with word processing and graphics software. For example, the way that the computer expects you to work is to select something first and then to issue a command that says what you want to do with whatever you have selected—either using keyboard commands or using the mouse, pointer, and menus. Once you have selected a region, for example, you can use the Cut, Copy, Paste, and Clear commands to rearrange and edit the material in your tracks.

You can select a region or regions using the Time Grabber tool or you can select a range along a track using the Selector tool. You can also work across multiple tracks. You can then use the Cut command to remove whatever you have selected from the Edit window and put this selection into the Clipboard ready to paste it elsewhere. The Clipboard is a temporary storage area in the computer's random access memory (RAM). You can use the Copy command to copy your selection into the Clipboard ready to paste elsewhere, without removing the original selection from the Edit window. You can use the Paste command to put the contents of the Clipboard into the Edit window at the Edit insertion point, overwriting any material that is already there. If you simply want to remove your selection without putting it into the Clipboard, use the Clear command instead.

Special Cut, Copy, Paste, and Clear Commands

Pro Tools also has four "special" Edit menu commands that you can use for editing automation playlists (e.g., volume, pan, mute, or plug-in automation) on audio, Auxiliary Input, Master Fader, and Instrument tracks.

Cut Special, Copy Special, and Clear Special each have three submenu selections. These let you edit all the Automation or just the Pan Automation (whether the data is showing in the track or not), or just the Plug-in Automation on its own (when this is showing in the track).

The Paste Special command also has three submenu selections, the most useful of which is the "Repeat to Fill Selection" command. This allows you to automatically fill a selection with audio regions or data much more quickly than by manually duplicating the regions.

> **Tip**
>
> To use the "Repeat to Fill Selection," simply cut or copy an audio region so that it is in the Pro Tools software's Clipboard (i.e., temporarily stored into RAM), then make a selection in the Edit window and use the command to fill this selection. The Batch Fades window automatically opens to let you apply crossfades between the pasted regions.

Repeat to Fill Selection

The Repeat to Fill Selection command lets you automatically fill a selection with audio or MIDI regions or data without requiring that you manually duplicate the regions. To use Repeat to Fill Selection, Cut or Copy a region, then make a selection and use the command to fill the selection.

When pasting audio regions, you are prompted to specify a crossfade to be used for the pasted regions. If you fill an area that is an exact multiple of the copied region size (e.g., filling 16 bars with a 4-bar region), the copied selection is pasted as many times as it takes to fill the selection.

If you fill an area that is not an exact multiple of the copied region size (for example, filling 15 seconds of a track with a 2-second region of room noise), the remaining selection area is filled with an automatically trimmed version of the original selection.

Stripping Silence from Regions

The Strip Silence feature lets you analyze audio selections containing one or more regions across one or more tracks to define areas that you wish to regard as "silence." Once you have identified these, you can "strip away" (i.e., remove) the "silence," or you can keep (i.e., "extract") the "silence" by removing the rest of the audio. The third option is to separate your selection into lots of smaller regions so that these can be quantized or edited individually afterwards.

Four sliding controls in the Strip Silence window let you set the parameters by which "silence" will be defined.

> **Tip**
>
> If you press Command (Mac) or Control (Windows) while adjusting the sliders you get finer resolution.

Figure 6.3 The Strip Silence window.

The Strip Threshold parameter lets you set a value for the amplitude, below which any audio is considered to be silence. For example, on a bass drum track, you will probably hear the sound of the rest of the drum kit when the bass drum is not playing, but this will be much quieter than the bass drum, in other words, the amplitude of this quieter audio will be much less than the amplitude of the bass drum sound. You can define this low amplitude audio to be considered as silence by setting the Strip Threshold just above this.

> ### Note
>
> In Pro Tools 8, Strip Silence lets you adjust the Strip Threshold down to –96 dB (the lower limit used to be –48 dB). This increased dynamic range is especially useful when working on recordings with low signal levels (such as ambient recordings) and recordings with a wide dynamic range (such as orchestral music).

You can also define what is to be considered to be silence using the Minimum Strip Duration parameter. This sets the minimum amount of time that the material below the threshold must last for before it is regarded as "wanted" audio.

The Region Start Pad parameter lets you extend the start of each new region to include any wanted audio material that falls below the threshold, such as breathy sounds before a vocal or the sound of fingers sliding up to a note or chord on a guitar. Similarly, the Region End Pad parameter lets you extend the end of each new region to make sure that the full decay of the audio material is preserved.

As you adjust the controls, rectangles start to appear in the selected region surrounding the wanted audio. The audio in between these rectangles is considered to be silence.

For example, take a look at the screenshot below. This shows a selection from the bass drum track of a multi-mic'ed drumkit. The bass drum inside the selection plays "bu-boom....bu-boom....boom....boom....bu-boom." These four sets of notes are outlined by four black rectangles. Everything in between is considered to be silence.

Figure 6.4 With the Strip Silence parameters correctly set, you see black rectangular boundaries around the "wanted" audio.

If you pad the start, this opens up the rectangles before the drum beat and if you pad the end, this opens up the rectangles after the drum beats—including more audio before and after the drum beats. As you can see in the accompanying screenshot, it is unable to do this before and after the region selection. To assess how much to pad the sounds, you will need to strip the silence, audition carefully, undo the strip silence and adjust the pad settings, strip the silence again and audition carefully again, and repeat this until you have the results you want. Listen carefully to the decay of the sound, for example. If you cut this too short, it can make the recording sound too unnatural.

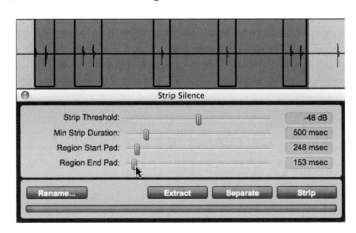

Figure 6.5 Adjusting the Start and End Pads.

When you click the Strip button, it removes the areas that you have defined as silence from the selected region. In the accompanying screenshot you will see the five "wanted" regions containing the bass drum notes. Now, you can move the bass drums around to change the timing or "feel"—or edit these whatever way you wish.

Figure 6.6 The silence between the "wanted" regions has been stripped away by clicking "Strip."

If you click the Extract button instead, this removes the audio above the Strip Threshold and leaves (or "extracts") the audio that you have defined as silence. So you can think of Extract as the inverse of Strip with the "wanted" audio, in this case, being the audio that falls below the Strip Threshold. This feature could be useful in postproduction if you wanted to extract the "room tone" or ambience from part of a recording to use elsewhere, for example.

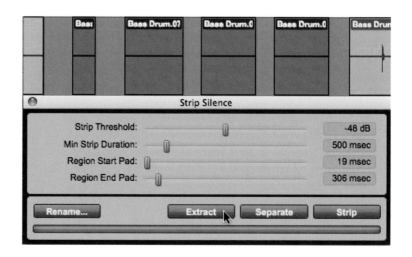

Figure 6.7 The audio above the Strip Threshold has been removed, leaving the audio defined as "silence," but which actually contains ambience, remaining (i.e., "extracted") in the track.

The third option is to click the Separate button. This automatically divides the selection into regions based on the boundaries detected by Strip Silence. In the accompanying screenshot, you will see that the selection (which I extended a little compared with the previous examples) has been divided into 10 separate regions, each with its own name. The Separate feature is very useful if you want to quantize the audio in each region to line up with the bars and beats, for example.

Figure 6.8 The Separate button divides the whole selection into separate regions based on the boundaries detected by Strip Silence.

If you click the Rename button in the Strip Silence window, this opens the Rename Selected Regions dialog. This lets you define how the regions that you will create using the Strip Silence feature will be named.

Whatever you type in the Name field will be used as the base name for the regions created using Strip Silence. For example, if you want a space to appear after the name and before the numbers, you will have to insert a space here. The Starting Number field lets you specify the number at which sequential autonumbering starts. The Number of Places field lets you specify the number of zeroes to use before the region numbers. You can also specify a Suffix to be used at the end of the name, after the number. You might use this to identify the "take" from which these regions have been created, for example.

For this example, I chose "Bass Drum" with a space after it as the Name; Starting Number 1; three as the Number of Places to give me two leading zeroes with each number up to 9, one leading zero before each number from 10–99, and no leading zeroes for numbers above 100. For my Suffix, I typed a space, followed by a hyphen, then another space, then "Take 1" to identify the take from which these bass drums had been stripped.

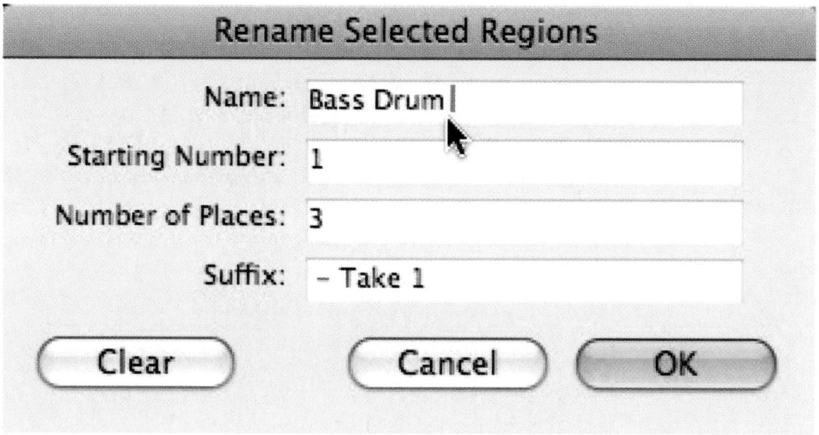

Rename Selected Regions

Name: | Bass Drum |

Starting Number: | 1

Number of Places: | 3

Suffix: | – Take 1

Clear   Cancel   OK

Figure 6.9
Rename Selected
Regions dialog.

Take a look at the accompanying screenshot to see how the regions were named in the Regions List as a result of this naming scheme.

Bass Drum 001 - Take 1
Bass Drum 002 - Take 1
Bass Drum 003 - Take 1
Bass Drum 004 - Take 1
Bass Drum 005 - Take 1
Bass Drum 006 - Take 1
Bass Drum 007 - Take 1
Bass Drum 008 - Take 1
Bass Drum 009 - Take 1
Bass Drum 010 - Take 1

Figure 6.10 Bass
drum regions
in Regions List
showing the
naming scheme.

Inserting Silence

The Insert Silence command does what it says, replacing a selection that you have made on a track or tracks with silence. Less obviously, it can also be used to remove automation data.

If the track is displaying audio or MIDI data, when you apply the Insert Silence command to a selected range, it not only clears the audio or MIDI data, it also clears any automation data for the track or tracks. However, if the selected tracks are displaying automation data, the automation data visible on each track is cleared throughout the selected range and any audio or MIDI is left untouched. Also, if you press the Control key (Mac) or Start key (Windows) while choosing the Insert Silence command, this clears all the automation data from all the selected tracks—not just the visible data.

The Insert Silence command is particularly clever when used in Shuffle mode. In this case, it moves the track data within and after the Edit selection forward, by an amount equal to the selection, pushing everything (including any auto-mation data) forward from the start of the selection, to get it "out of the way" of, thus making room for, the silence that you are inserting. To explain this another way, when you apply the Insert Silence command to an Edit selection while in Shuffle mode, Pro Tools splits the region, or regions, at the beginning of the insertion point and moves the new region, or regions, to a position later in the track by an amount equal to the length of the selection, before inserting the selected amount of silence.

Time Compression and Expansion (TCE) Edit to Timeline Selection

When the Edit and Timeline selections are unlinked, you can compress or expand an audio selection to fit the Timeline selection using the Edit menu's "TCE Edit to Timeline Selection" command. If you prefer a keyboard command, you can use Option-Shift-U (Mac) or Alt-Shift-U (Windows).

TCE Edit to Timeline Selection ⌥⇧U

Figure 6.11 Edit menu TCE Edit to Timeline Selection command.

The TCE Edit to Timeline command can be used on multichannel selections and selections across multiple tracks. All regions are compressed or expanded equally by the same percentage value, based on Edit selection range. This ensures that the rhythmic relationship between the different channels or tracks is maintained.

On Audio tracks, the "TCE Edit to Timeline Selection" command uses which-ever TCE AudioSuite Plug-in you have selected in the Processing Preferences. On Elastic Audio-enabled tracks, TCE Edit To Timeline Selection uses the track's selected Elastic Audio plug-in.

Figure 6.12
Selecting the TC/E
Plug-in preference.

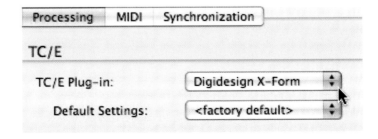

> ## Tip
>
> The standard Pro Tools TC/E plug-in sounds terrible and the Pro Tools Time Shift plug-in is only a little better than this. The Elastic Audio real time plug-ins are not very good either. These plug-ins are only usable to give you a rough idea of how things will sound as you are working ideas out. You will need to buy a much higher quality plug-in if you want to achieve professional results. I can recommend Serato's Pitch 'n Time, which has very professional features. Avid also offers an excellent Pro Tools plug-in for time compression and expansion called X-Form that I am currently using. This takes quite a lot longer to process the audio, but delivers truly professional results.

To fit an Edit selection to the Timeline, Disable "Link Timeline and Edit Selection." With the Selector tool, select the audio material to be compressed or expanded. In any Timebase ruler, select the time range where you want to fit the audio material then choose "TCE Edit to Timeline Selection" from the Edit menu.

Figure 6.13 An Edit selection ready to be fitted to a Timeline selection.

The Edit selection is compressed or expanded to the length of the Timeline selection.

Figure 6.14 The Edit selection has been compressed and moved to fit into the Timeline selection using the "TCE Edit to Timeline Selection" command.

Consolidating Regions

Often, during your editing sessions, you will end up with tracks made up from many smaller regions. This could happen if you "comp" several takes to make a composite containing the best from each take. When you are satisfied with your edits to these tracks, Pro Tools allows you to consolidate a track, or a range within a track (such as a verse or chorus), into a single region, which is much easier to work with. When working with audio tracks, this consolidation process causes a new audio file to be created that encompasses the selection range, including any blank space, treating any muted regions as silence.

To use this feature, first you select the regions you want to consolidate using the Time Grabber tool or the Selector tool. A quick way to select all the regions in a track is to triple-click in its playlist with the Selector tool. When you choose Consolidate from the Edit menu, this will create a new, single region that replaces the previously selected regions. If you prefer to use a keyboard command instead, you can press Option-Shift-3 (Mac) or Alt-Shift-3 (Windows).

> **Tip**
> Consolidating an audio track does not apply any automation data as it creates the new file, so if you want to create a new file with automation data applied to the audio, use the "Bounce to Disk" feature instead.

The Edit Modes

Pro Tools has four Edit modes—Slip, Shuffle, Spot, and Grid—that can be selected by clicking the corresponding button in the upper left of the Edit window. You can also use the function keys on your computer keyboard, F1 (Shuffle), F2 (Slip), F3 (Spot), and F4 (Grid), to select the mode. The Edit mode that you have chosen affects the ways that regions may be moved or placed, how commands like Copy and Paste work, and how the various Edit tools (Trim, Selector, Grabber, and Pencil) work.

Slip Mode

The default mode that you should normally work in is the Slip mode. It is called Slip mode because it lets you move regions freely within a track or to other tracks and allows you to place a region so that there is space between it and other regions in a track.

In Slip mode, if you use the Cut command to remove a selection before the end of the last region on a track, it leaves an empty space where the data was removed from the track. You should also be aware that regions are allowed to overlap or completely cover other regions in Slip mode.

Shuffle Mode

In Shuffle mode, if you place two or more regions into a track, they will automatically snap together with no gap in between.

If you have an existing track that contains two or more regions with gaps between them, you can close the gaps by selecting Shuffle mode and using the Time Grabber tool to push a region in the direction of the previous region.

Figure 6.15
Selecting and moving a region in Shuffle mode.

As soon as you let go of the mouse, the gap between the two regions is closed, leaving them "stuck together."

Figure 6.16 The region that you moved automatically attaches itself to the previous region in Shuffle mode.

So, if you want a region that you are moving around in the Edit window to automatically butt up against the previous region, with no overlap and with not even the smallest gap between them, you can use Shuffle mode. If you are in Slip mode or even in Grid mode, you have to be careful to avoid overlapping the regions and you will have to zoom in to see what you are doing. Consequently, placing regions accurately in these modes takes more time.

By the way, it is called "Shuffle" mode because if you use the grabber to drag a region placed earlier in a playlist to a later position (or vice versa), the other regions will shuffle (i.e., move) their positions around to accommodate this repositioning of the region. Similarly, if you use the Cut command to remove a selection before the end of the last region on a track, the regions to the right of the cut move to the left, closing the gap. Also, if you paste data anywhere before the end of the last region on a track, all regions beyond the insertion point move to the right to make room for the pasted material. And if you Paste in Shuffle Mode, the audio at the edit point gets shifted to the right, unless a range is selected, in which case this range gets completely replaced.

> **Tip**
>
> Be careful to return to Slip or Grid mode as soon as you have made your moves in Shuffle mode—it is all too easy to accidentally move a region and have Shuffle mode shuffle your regions to somewhere they shouldn't be. And if you don't notice this at the time it happens, you may not be able to use even the multiple Undo feature to get back to where you were.

Spot Mode

Spot mode was originally designed for working to picture, where you often need to "spot" a sound effect or a music cue to a particular SMPTE time code location.

The way this works is that you select Spot mode and then click on any region in the Edit window, or drag a region from the Region List or from a DigiBase browser into the Edit window. The Spot dialog comes up and you can either type in the location you want or use the region's time stamp locations for spotting.

There are actually two time stamps that are saved with every region. When you originally record a region, it is permanently time-stamped relative to the SMPTE start time specified for the session. Each region can also have a User Time Stamp that can be altered whenever you like using the Time Stamp Selected command in the Regions List pop-up menu. If you have not specifically set a User Time Stamp, the Original Time Stamp location will be set here as the default.

> **Tip**
>
> Spot mode can also be very useful when editing music projects in Pro Tools—particularly if you move a region out of place accidentally. Using Spot mode you can always return a region to where it was originally recorded. Also, as long as you remember to set a User Time Stamp if you rearrange regions to locations other than where they were first recorded, you can always return regions to these locations.

For example, if you move a region by accident, using Spot mode would be an ideal way to put it back to exactly where it came from.

Figure 6.17 Edit window showing a region accidentally moved from its original position.

To do this, put the software into Spot mode by clicking on the Spot mode icon at the top left of the Edit window, select the Grabber tool and click on the region to bring up the Spot dialog.

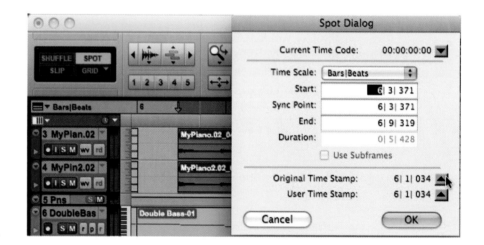

Figure 6.18 Using the Spot Dialog to return a region to its original position.

If the Region has been moved from the position at which it was originally recorded, the current Start position displayed in the Spot dialog will be different from the Original Time Stamp position shown in the dialog. You can either type the correct Start position or (even quicker) just click the upwards-pointing arrow next to the Original Time Stamp to put this value into the Start position field.

Figure 6.19 Click the Original Time Stamp arrow to enter this location into the Start field in the Spot Dialog.

Note

If you have deliberately moved the Region since recording it and have taken the trouble to enter a User Time Stamp in the Regions List for the moved region, you should enter the User Time Stamp into the Start field instead. If you did not update the User Time Stamp when you moved the region, then you will not know what the correct location should be, so you will not be able to return the region to the correct position using Spot mode.

When you click "OK" in the Spot dialog, the region will be moved back to the location where it was originally recorded.

Figure 6.20 Edit window showing the "MyPiano" Region back in its original position.

Grid Mode

Grid mode lets you constrain your edit selection to gridlines that correspond to a grid value that you can choose to suit your purpose. Grid mode is particularly useful if you are editing pattern-based music that starts and ends cleanly at regular boundaries, such as bars or beats. You can choose the Grid size using the Grid value pop-up menu located above the Timebase rulers and the tracks in the Edit window.

Tip

If you press and hold the Control and Option keys (Mac) or the Start and Alt keys (in Windows) you can use the plus (+) and minus (−) keys on the numeric keypad to increment or decrement the Grid size.

The Grid size can be based on a time value using the Main Time Scale, or, if Follow Main Time Scale is deselected, another time format can be used for the Grid size. To make the Grid lines visible in the Edit window, just click on the currently selected Timebase ruler name (the one highlighted in blue) to toggle these on and off.

Tip

You can temporarily suspend Grid mode and switch to Slip mode by holding down the Command key (the Control key in Windows), which is very useful while you are trimming regions, for example.

Absolute and Relative Grid Modes

Grid mode can be applied either in an Absolute or Relative way. In Absolute Grid mode, regions are "snapped" onto Grid boundaries when you move them—so regions can never be placed in between the currently applicable Grid boundaries. This is what you normally expect and want Grid mode to do and is the default behaviour.

Relative Grid Mode lets you edit regions that are not aligned with Grid boundaries as though they were. For example, in 4/4 time signature, if a region's start point falls between beats and the Grid is set to 1/4 notes, dragging the region in Relative Grid mode will preserve the region's position relative to the nearest beat.

In a music recording, for example, a musician may have deliberately played a note just before the beginning of a new bar. If you separate this note into its own region, then use the Grabber tool to move it earlier or later by an exact number of beats using the Absolute Grid mode, the region's start position will be "snapped" exactly onto the new beat position, and will not sound as the musician intended it to.

If you choose Relative Grid mode and set a suitable Grid value that allows sufficient time before the bar, then when you move the region containing the note along the Grid, it "snaps" to positions that preserve the note's original positioning just a little before the beat.

To select Relative Grid mode, click and hold the Grid button, move the mouse to where it says "Relative Grid," then let go.

Figure 6.21
Selecting Relative Grid mode.

As you can see from the accompanying screenshot, the bass plays four "pickup" notes before the beginning of bar 2. If you want to drag this whole region to start at the next bar with Grid mode set to Bars, using Absolute Grid mode it would snap the beginning of the region to the next bar line. This would make no sense because the "pickup" notes need to play before the bar line.

Figure 6.22 The bass has some pickup notes.

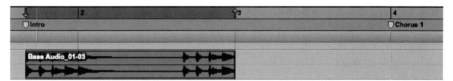

With Relative Grid mode selected, when you use the Grabber tool to slide the region to right or left along the Timeline, the region snaps to start positions that are exactly the same distance before the new bar position as those in the original region, thus preserving the note's original positioning relative to the bar positions.

In the example below, I dragged the region later by one bar in Relative Grid Mode and as you can see from the screenshot, it is now positioned a little before Bar 3.

Figure 6.23 Note dragged along the Timeline in Relative Grid mode to the correct position before Bar 3.

Dragging the region in Absolute Grid mode would have only allowed me to position the start of the region containing the note exactly onto a sixty-fourth note. When I do this, as you will see from the accompanying screenshot, the note starts exactly on a sixty-fourth note grid position—which in this instance would sound wrong.

Figure 6.24 Note dragged back along the Timeline in Absolute Grid mode to a new bar position.

Editing in the Edit Window

The Pro Tools Edit window lets you edit audio nondestructively—editing a visual representation of the audio data without altering the audio source files. The regions that represent the audio can be cut, copied, pasted, trimmed, or cleared from the Edit window with no risk that you will permanently alter or lose any of your precious audio recordings.

There are some tools and processes that do work destructively, in other words, that can permanently change audio files on your hard disk. One important example is when you use the pencil tool to redraw an audio waveform, which does alter the audio file on disk when you save your changes. Another example is when you use the AudioSuite processes—but these are normally accompanied by clear warnings to this effect.

One of the great things about Pro Tools is that you can carry out many editing tasks while the session is playing back. You can separate and trim regions, place, spot or rearrange regions, add fades or crossfades, transpose or quantize MIDI tracks, nudge audio or MIDI regions, listen to different playlists, insert real-time plug-ins, and, perhaps most usefully, edit automation breakpoints. This makes it extremely fast to edit sessions—which can be a crucial factor on busy commercial recording sessions.

Whenever you record or import audio, video, or MIDI data to tracks in Pro Tools, regions are created that represent this data visually in the tracks. Regions indicate where the material begins and ends and also give useful visual feedback about the general shape and content. New regions are created when you resize or separate, cut or paste existing regions, and all regions of all types are displayed in the Regions List from which they can be dragged into tracks in the Edit window.

> **Tip**
>
> A region can represent a complete audio file or just part of an audio file. For example, if you import a track from a CD, a region is created that represents the whole track. If you then edit this, say, by trimming the region in the Edit window, a second region will be created that represents less than (by the amount you trimmed) the whole track.

Edit Window Track Views

By default, the Edit window is set to show tracks in "Waveform" view. This means that you see a representation of the audio waveform presented as a graph of amplitude versus time. Audio tracks can also be set to show Playlists, Blocks, Volume, Volume Trim, Pan, Mute, or any plug-in controls that have been enabled for automation. You can also select the Elastic Audio Analysis and Warp views here. The default view for Audio tracks is Waveform view.

Auxiliary inputs can be set to Volume, Volume Trim, Pan, Mute, or plug-in automation controls; Master Fader Tracks can be set to Volume, Volume Trim, or plug-in automation controls; VCA Master Tracks can be set to Volume, Volume Trim, or Mute. The default view for VCA, Auxiliary, and Master tracks is Volume automation view.

MIDI tracks are usually set either to Notes or to Regions and can also be set to Blocks, Volume, Pan, Mute, Velocity, Pitch Bend, After Touch, Program, Sysex, or any continuous controller type. Instrument tracks, like MIDI tracks, are usually set either to Notes or to Regions, and can also be set to Blocks, Volume, Pan, Mute, Velocity, Pitch Bend, After Touch, Program, Sysex, or any continuous controller type as well as Volume, Volume Trim, Pan, Mute, or any plug-in controls that have been enabled for automation.

The default view for MIDI and Instrument tracks is Regions view, which is the most useful view for editing and arranging MIDI regions in the Edit window.

> **Tip**
>
> Block view can be useful when you are cutting and pasting regions and want to move them around to rearrange them.

> **Note**
>
> For detailed MIDI editing, you can double-click any MIDI region to open it in a MIDI Editor window.

Master Views

Audio tracks have two Master Views: Waveform and Blocks; MIDI and Instrument tracks have three: Regions, Blocks, and Notes (when using the Selector tool); and Auxiliary Input, Master Fader, and VCA Master tracks just have one: Volume.

When a track is in a Master View, any edits that you perform apply to all data in the track—including all the automation data. If you are working with data from multiple tracks and any of the selected tracks is set to a Master View, any edits you make will not only affect the audio or MIDI data on the track, they will also apply to any automation or controller data on the track. So, when you are in a Master View, cutting an audio region also cuts any volume, pan, mute, send, or plug-in automation that is also on the track—saving you from having to individually cut data from each automation playlist on the track. Similarly, when an Auxiliary Input or Master Fader track is displayed in its Master View, any edits performed apply to all automation data in the track.

> ## Note
>
> When Audio tracks are set to display either Waveform or Blocks, they are said to be in a Master View. When MIDI or Instrument tracks are set to display Regions, Blocks, or Notes (when using the Selector tool), these tracks are said to be in a Master View. When Auxiliary inputs are set to display Volume, this is regarded as the Master View, and Master Fader tracks (which can only display Volume) are always considered to be in Master View.

In any other track view with automation data displayed on the selected tracks, edits only affect the type of automation data displayed in each track. For example, if track 1 displays Pan automation, track 2 displays Volume automation, and track 3 displays Mute automation, the Cut command cuts only pan data from track 1, volume data from track 2, and mute data from track 3.

> ## Tip
>
> Pro Tools lets you override this behaviour temporarily by pressing and holding the Control (Mac) or the Start key (Windows) while you choose the Cut, Copy, or Paste commands, enabling you to copy all types of automation on all selected tracks.

Breakpoints

Breakpoints are points that can be inserted onto the graph lines that control automation data in the Pro Tools Edit window. As the manual explains "When an audio or Instrument track is displayed as Volume, Pan, or another automated control, or when a MIDI or Instrument track is set to one of the continuous controller types (Volume, Pitch Bend, After Touch), the data for that track appears in the form of a line graph with a series of editable breakpoints. The breakpoints can be dragged to modify the automation data, and new breakpoints can be inserted with the Pencil tool or a Grabber tool." Got that?

All I would like to add to this is that editing volume and other automation data can often be done much more quickly, and much more accurately, by editing the breakpoints than by moving faders and other controls—which tends to generate lots of unnecessary automation data. Of course, if you like the feel of faders, knobs, and switches, nothing is going to substitute for these!

Switching Track Views

You can switch the view of any track using its Track View pop-up menu. When you have the Track Height set to Medium or greater, this is located to the left of the track in the Track Controls display area, just underneath the Record, Solo, and Mute buttons. Otherwise, you can select this from the Track Options pop-up menu at the top left of the Controls area.

Tip

Use keyboard command shortcuts to get to the views you want. To change to the previous or next Track View: Click in the track you want to change. If this track is a member of a group, all the other members will be selected. If you want to change views on more than one track, Shift-click or drag the Selector tool to select additional tracks.

With the track or tracks selected, just press and hold the Control and Command keys (Mac) or Control and Start keys (Windows); with these keys held down, you can use the Left or Right Arrow keys on your computer keyboard to select the previous or next Track View.

To change the view between Waveform and Volume (audio) or Notes and Regions (MIDI): Just select the relevant track or tracks, as above, then press the Control and Minus (Mac) or Start and Minus (Windows) keys on your computer keyboard.

Useful Keyboard Commands for Editing

Pro Tools provides a useful keyboard command that lets you "collapse" an Edit (or Timeline) selection and simultaneously move the Edit cursor insertion point to the beginning or end of the selection. Pro Tools also provides several keyboard shortcuts for moving and extending or decreasing the range of an Edit (or Timeline) selection. The best way to learn how these work is to make a selection, as I have in the accompanying screenshots, and try these for yourself.

With the Transport stopped, you can use the Down and Up arrows on your computer keyboard to place the Edit cursor at the beginning or end of the Edit (or Timeline) selection and collapse the selection.

Figure 6.25
A region selected in the Edit window with Timeline Selection Start and End Markers shown in red in the Bars|Beats ruler (because a track happens to be record-enabled).

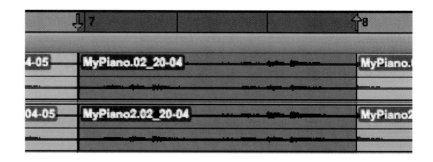

To move the insertion point to the beginning of the selection, press the Down arrow, or to move the insertion point to the end of the selection, press the Up arrow. You will "lose" the selection, which is then said to have "collapsed," and the insertion point will appear at the beginning or end of the selection.

Figure 6.26
Selection "collapsed," with the Edit cursor inserted at the end of the selection and the Timeline Selection Marker placed above this in the Bars|Beats ruler.

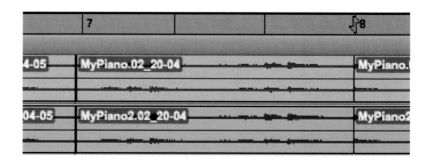

To halve the length of the selection, press Command-Control-Option-Shift-L (Mac) or Control-Alt-Start-Shift-L (Windows).

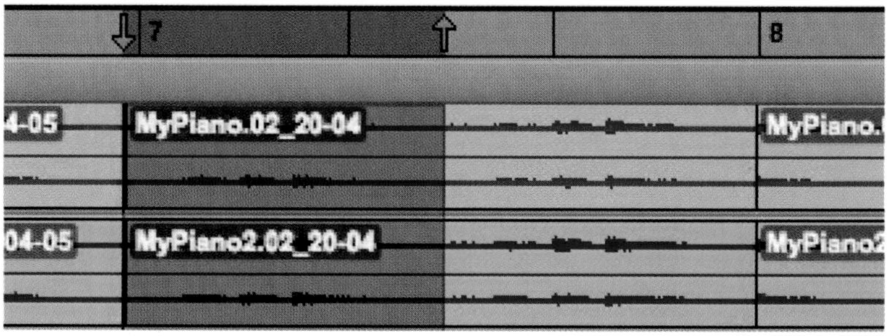

Figure 6.27 The Selection has been halved.

To move the selection forward by the selection amount, press Command-Control-Option-' (single quote) (Mac) or Control-Alt-Start-' (single quote) (Windows).

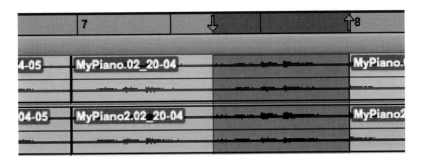

Figure 6.28 The Selection has been moved forward by the selection amount.

To move the selection backward by the selection amount, press Command-Control-Option-L (Mac) or Control-Alt-Start-L (Windows).

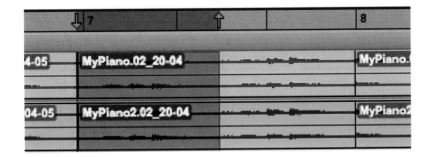

Figure 6.29 The Selection has been moved back again.

To double length of the Edit selection, press Command-Control-Option-Shift-' (single quote) (Mac) or Control-Alt-Start-Shift-' (single quote) (Windows).

Figure 6.30 The Selection has been doubled.

To align the start point of a region to timeline position, select the Grabber tool and use the mouse to point at and click in the timeline ruler to position the Edit cursor at exactly the point at which you wish to align the start point of the region. Then, while holding down the Control key (Mac) or Start key (Windows), click on the region and watch it snap exactly to the timeline position that you have specified!

> **Note**
>
> This trick even works when you are dragging regions into the timeline from the Regions list—and can be faster to use than Spot mode (which lets you do much the same thing).

To align the end point of a region to timeline position, select the Grabber tool and use the mouse to point at and click in the timeline ruler to position the Edit cursor at exactly the point at which you wish to align the end point of the region. Then, while holding down the Command and Control keys (Mac) or Control and Start keys (Windows), click on the region and watch it snap exactly to the timeline position that you have specified.

> **Note**
>
> This can be very useful if you want to align two sound files whose end point is the same but whose start points are different.

To align the sync point of a region to timeline position, select the Grabber tool and use the mouse to point at and click in the timeline ruler to position the Edit cursor at exactly the point at which you wish to align the sync point of the region. Then, while holding down the Control and Shift keys (Mac) or Start and Shift keys (Windows), click on the region and watch it snap exactly to the timeline position that you have specified.

> **Note**
>
> This is particularly useful for aligning sound effects that have a sync point somewhere other than at the start of the region, such as explosions or door slams.

Working with Regions

Regions are the basic building blocks for arranging audio and MIDI in the Edit window, so you need to make sure that you know how these are created, edited, and arranged.

When you record new audio or import existing audio files, Pro Tools creates a region that plays back the entire file when placed into a playlist on a track. Very often you will use the Trimmer tool to remove some audio from the start or end of the recording. This creates a new region in the playlist that is shorter than the original region, and this new region also appears in the Region List.

New regions are often created automatically when you make edits to existing regions. For example, if you use the Clear command to remove a section from within a region, the sections on either side of this will form new regions. If you use the Cut command instead, the section you are cutting is placed onto the Clipboard and a new region representing this section is also added to the Region List.

Once you have created a region, it appears in the Region List. From the Region List, you can drag it to a track to add to an existing arrangement of regions or to create a new arrangement "from scratch." You can slide regions or groups of regions around freely in the Edit window using the Time Grabber tool in Slip mode. You can also move regions around while constrained to the grid in Grid mode, or shuffle them around in Shuffle mode, or "spot" regions to exact locations using Spot mode.

Capturing Regions

The "Capture…" region command defines a selection as a new region and adds it to the Region List. From there, the new region can be dragged to any existing tracks.

Figure 6.31 A selection within a region.

When you make a selection within a region and choose "Capture…" from the Region menu or press Command-R (Mac) or Control-R (Windows), the Name dialog appears. You can type a name for the captured region here and OK this.

Figure 6.32
Naming a selection within a region using the Region menu's Capture… command.

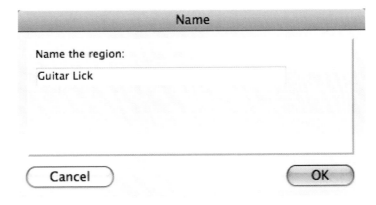

269

This captured region then appears in the Regions List from where it can be dragged and dropped into any track in the Edit window at any location.

Figure 6.33 The captured Guitar Lick region can be dragged from the Regions List and placed anywhere in the Edit window.

Separating Regions

You can separate regions with the Selector tool enabled by inserting the cursor anywhere in a region and pressing Command-E (Mac) or Control-E (Windows) or by choosing "Separate Region: At Selection" from the Edit menu. This simply cuts the region into two at the insertion point. If you make an Edit selection within a region and use this command to separate the region, this will create two separate regions, or three if you leave parts of the original region unselected either side of your selection.

If you need to separate a region into lots of smaller regions, you can use one or other of the two additional choices available in the Separate Region submenu: "Separate Region On Grid" separates regions based on the currently displayed Grid values and boundaries. "Separate Region At Transients" separates regions at each detected transient.

Using either of these commands opens a "Pre-Separate Amount" dialog that you can use to type a preseparate amount in milliseconds to pad the beginnings of the new regions with a small extra amount of audio.

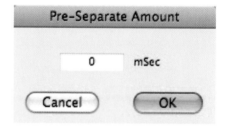

Figure 6.34 Pre-Separate Amount dialog.

> **Note**
>
> With looped regions, the Separate Regions commands automatically unloop and flatten the looped regions before separating them.

The Auto-Name Separated Regions option in the Editing Preferences page is selected by default, so Pro Tools automatically names separated regions for you.

If this preference is disabled, a dialog box appears that lets you type a name for the new region.

Figure 6.35
Editing Preferences
for Regions.

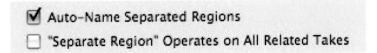

☑ Auto-Name Separated Regions

☐ "Separate Region" Operates on All Related Takes

Note

When the "Separate Region" Operates on All Related Takes option is enabled in the Editing Preferences, editing a region with the Separate Region command also affects all other related takes with the same User Time Stamp. This option helps you compare different sections from a group of related takes. For example, you can quickly separate an entire group of related vocal takes into sections, then audition and select the best material from each section independently. If this option is selected, make sure the Track Name and Region Start and End options are also selected in the Matching Criteria window. If they are not, *all* regions in the session with the same User Time Stamp will be affected. In most instances, you will want to disable this option, to prevent a large number of regions from being created when you use the Separate Region command.

Using the Separation Grabber

You can also create new regions using the Separation Grabber tool. Using the Separation Grabber saves you the trouble of separating the region first. Make a selection using the Selector tool. Then select the Separation Grabber tool from the pop-up selector that appears when you press and hold the Grabber tool's button. Take a look at the accompanying screenshot to see how this might look.

Figure 6.36
Choosing the
Separation
Grabber tool.

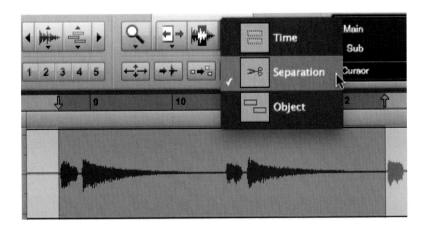

When you have selected the Separation Grabber tool, you will see a small pair of scissors appear on the "hand" to remind you that the Separation Grabber is currently selected. Now, you can drag your selection to a new location within the same track or on another track. The selection is automatically separated from its "parent" region, and a new region containing this selection is created. As usual, new regions are also created from the material outside the original selection. Take a look at the accompanying screenshot to see how this might look.

Figure 6.37
Using the Separation grabber tool to separate a selection from a region and move this new region to a new location.

Tip

If you want to leave the original region intact, press the Option key (Mac) or the Alt key (Windows) while you drag the selection to the new location. A new region containing a copy of the previous selection is created and placed at the new location while the original region is not moved or changed in any way.

Healing Separated Regions

If you have made a cut in a region or removed a section from within a region, you may change your mind about this and want to return it to its original condition. If the regions have not been moved around since you made the cut or removed material, you can join the separated regions back together again using the Heal Separation command. The Heal Separation command returns separated regions to their original state—provided that the regions are still next to each other and that their relative start/end points haven't changed since they were separated.

To heal a separation between two regions, use the Selector tool to make a selection that includes part of the first region, the entire separation between the regions, and part of the second region. Then, choose Heal Separation from the Edit menu or press Command-H (Mac) or Control-H (Windows) to heal the separation.

Note

You cannot join two regions created from different audio files together using Heal Separation because they were never part of the same region previously.

Overlapping Regions

Sometimes when you move regions around in Slip mode, you will inadvertently overlap these. You will also get overlapping regions on tick-based tracks with multiple regions if you increase the tempo. To warn you that this has happened, you can choose to display a small "dog-ear" in the top right-hand corner of each region if an overlap (or an "underlap") has occurred. Just select "Overlap" view for regions in the View menu.

The Regions menu also provides a couple of useful commands that let you bring an overlapped region to the front or send it behind neighboring regions as necessary. If you select multiple regions and choose "Bring to Front" from the Region menu or press Option-Shift-F (Mac) or Alt-Shift-F (Windows), the first region is placed in front of the second region, the second region is placed in front of the third region, and so forth. Take a look at the accompanying screenshot to see how this looks.

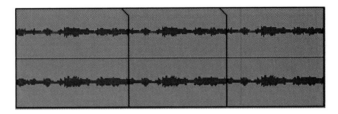

Figure 6.38 Three selected overlapping regions, each brought to the front before the next region, so each overlaps the next.

If you select multiple regions and choose "Send to Back" from the Region menu or press Option-Shift-B (Mac) or Alt-Shift-B (Windows), the second region overlaps the first region, the third region overlaps the second region, and so forth. Again, take a look at the accompanying screenshot to see how this looks. To describe this situation another way, you could say that each region "underlaps" the following region.

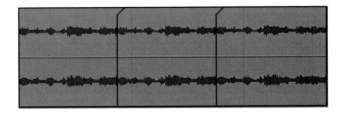

Figure 6.39 Three selected overlapping regions, each sent to the back before the next region, so each "underlaps" the next.

Now, if you select the middle of the three regions and send this to the back, the region before and the region after will both overlap this.

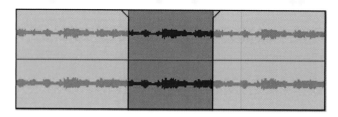

Figure 6.40 The middle of three regions has been sent to the back so that the region before and the region after both overlap this region.

And if you select the middle of the three regions bring this to the front, it will overlap both the region before and the region after.

Figure 6.41 The middle of three regions has been brought to the front so that It overlaps both the region before and the region after.

The easiest way to understand how all this works is to look at the accompanying screenshots, then try it for yourself.

Sync Points

When you are placing regions in Spot mode or Grid mode, it is sometimes useful to align a particular point within a region with a specific Timeline location—instead of aligning the start point of the region, which is the default situation. To cater for such situations, Pro Tools lets you define a sync point for any region.

This situation often comes up when you are laying up music and sound effects to picture. The standard example quoted here is where you have a creaking door that eventually slams shut. The sound effect file includes the creak and the slam, and the obvious thing to do here is to line up the sound of the door slamming shut with the video frame at which the door actually slams shut—which is some way into the audio region representing this sound effect.

You can define a region's sync point in Slip mode by first placing the Selector tool's insertion cursor at exactly the point within the region where the sound that you want to synchronize to start. Then choose the Identify Sync Point command from the Region menu to identify this as the sync point for the region.

Figure 6.42 Setting a sync point within a region with the Selector tool's insertion cursor placed at the position within the region where the sync point is to be placed.

A small down arrow appears at the bottom of the region, with a vertical, light gray line above this indicating the location of the sync point.

Figure 6.43 The sync point appears within the region as a small down arrow with a vertical, light gray line above this.

You can move this sync point to anywhere else inside the region using the Time Grabber tool to drag it earlier or later. To remove a sync point, just Option-click (Mac) or Alt-click (Windows) on the sync point using the Time Grabber or Scrubber tools.

> **Tip**
>
> Hearing the audio scrubbing back and forth can often be far more revealing than looking at the waveform when you are trying to identify what sounds best. So, you may find it helpful to use the Scrubber tool to help you locate where the sync point should be.

Nudging Regions

Pro Tools lets you nudge regions (or MIDI notes) by small increments or decrements along the tracks in the Edit window. The way this works is that you set a Nudge value using the Nudge Value pop-up menu, then select a region or group of regions; then you move (i.e., nudge) these forwards or backwards along the Timeline by pressing the plus (+) or minus (–) keys on the numeric keypad.

You can nudge material while Pro Tools is playing back, which really helps when you are fine-tuning "grooves." You can nudge continuously in real time to adjust the timing relationship between tracks and you can even nudge the positions of automation breakpoints in the playlists.

The Nudge value not only determines how far regions and selections are moved when you press the nudge keys, it can also be used to move the Start and end points for selections by the Nudge value or to trim regions by the Nudge value.

Figure 6.44 Setting the Nudge Value.

A pop-up selector near the top of the Edit window lets you choose the Nudge value. With the main counter set to Bars:Beats, for example, the values offered are the common subdivisions of a bar. You can also type the values you want directly into the Nudge value display, which is useful if the values you want to use are not listed in the pop-up.

> **Tip**
>
> If you press and hold the Command and Option keys (Mac) or the Control and Alt keys (in Windows) you can use the plus (+) and minus (–) keys on the numeric keypad to increment or decrement the Nudge value.

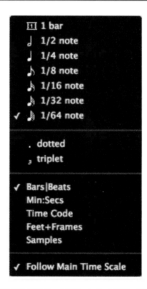

Figure 6.45 The Nudge value pop-up.

> **Tip**
>
> You can also nudge a region's contents (sliding audio or MIDI into and out of the current region boundaries) while keeping the region's start and end points exactly where they are—assuming that there is material outside the region's start and end points that can be slid into or out of the region. Using the Time Grabber tool, select a region whose contents you want to nudge. Press and hold the Control key (Mac) or the Start key (Windows) while you use the plus (+) and minus (–) keys to nudge the contents of the region without changing the region's start and end points.

Right-click Commands for Regions

A really useful technique for editors is to use Right-click commands with key combinations to perform operations on objects while preserving selections in the Edit and Mix windows.

For example, after you have made a region selection in the Timeline, or a Region name selection in the Region list, or a Track selection in the Edit or Mix window, you may wish to perform another operation without "losing" this selection.

To preserve your selection while performing another operation in these circumstances, Command-Right-click (Mac) or Control-Right-click (Windows) the object and choose a command from the pop-up "context" menu that appears.

(A different menu will appear depending on the "context," i.e., on which object you are clicking.)

You can use this technique to open one region's Right-click menu while you have another region selected, for example.

Figure 6.46
As can be seen here, it is possible to preserve the selection of one region (e.g., dub v1_17) while opening the Right-Click Menu of another region (e.g., Gtr_theme).

Or you can keep one track selected while carrying out an operation on another track, such as opening the Track Name Right-click menu—as can be seen in the accompanying screenshot.

Figure 6.47
While keeping the Vocal track selected (with Its name highlighted) It Is possible to open the Track Name Right-Click Menu for a different track (e.g., Guitartheme).

Region Editing Commands

The Edit menu and the Region menu both contain useful commands that you can apply to regions.

Shift Region

If you simply want to move selected material forward or backward in time along a track, and you know exactly how many bars and beats, seconds or samples, that you want to move this by, the quickest way is probably to use the Shift command. The Shift command works with selections, regions, MIDI notes, MIDI controller data, and automation breakpoints—and the selected material can reside on multiple tracks. To use this command, first select the track material you want to shift using the Selector or Time Grabber tool, then choose Shift from the Edit menu or press Option-H (Mac) or Alt-H (Windows). The Shift dialog that opens lets you choose whether the data will be moved earlier or later, and you can type the amount in bars and beats or using any of the other timebases.

> **Note**
> If you have selected just part of a region, when you shift it, new regions are created from the selection and from any material outside of the selection.

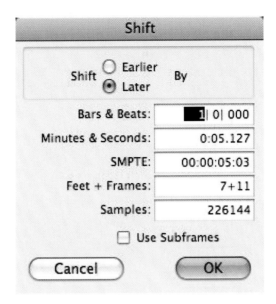

Figure 6.48 The
Shift dialog.

Duplicate Region

The Edit menu's Duplicate command copies a selection and places it immediately after the end of the selection. Though this is similar to using Copy and Paste, Duplicate is more convenient and faster, particularly when working with data on multiple tracks.

To duplicate a 1-bar or 2-bar (or other even bar number) selection when you are working in Bars|Beats so that it follows the original selection, use Grid Edit mode and the duplicate will be placed immediately after the selection's end point. The keyboard command to use for this is Command-D (Mac) or Control-D (Windows).

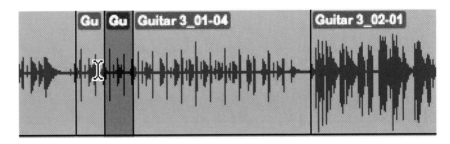

Figure 6.49 Using the Duplicate command: select a region using the Selector tool, then choose Duplicate from the Edit menu, or Use the keyboard command.

> ## Note
>
> When using Duplicate (or Repeat) for audio that must fall cleanly on the beat (such as rhythmic loops), it is important that you select the audio material with the Selector tool, or by typing in the start and end points in the Event Edit area. If you select an audio region with the Time Grabber tool (or by double-clicking it with the Selector tool), the material may drift by several ticks because of sample rounding.

To copy a region and automatically place it immediately before the selected region, choose the Time Grabber tool then click on the region while holding all three modifier keys: Command-Option-Control (on the Mac) or Start-Alt-Control (in Windows). This command actually creates a copy of the region and slides it earlier—it doesn't use the Duplicate command.

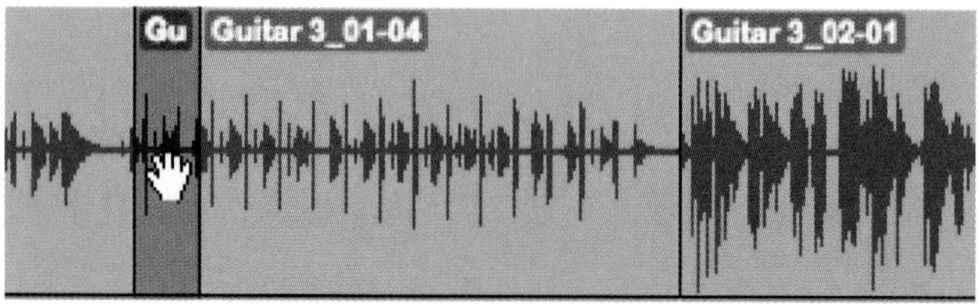

Figure 6.50 To duplicate a region and place it before the selected region, click using the three modifier keys.

Figure 6.51 The duplicated region is placed immediately before the region that was clicked on.

And it gets better! If you click on any other region in the track first to select it, then click on the region you want to duplicate while holding the three modifier keys, a duplicate of this region will attach itself immediately before the region you first selected.

Figure 6.52 Select a region before which you wish to place another region so that it goes dark.

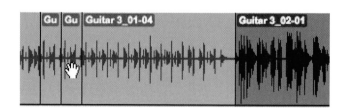

To carry out this move, select a region before which you wish to place another region so that it goes dark, then click on the region that you wish to place before the selected region using the three modifier keys. This command doesn't actually use the Duplicate command: it creates a copy of the region and slides it into position immediately before the previously selected region.

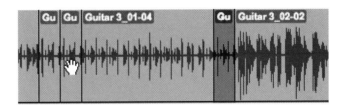

Figure 6.53 Click on the region that you wish to place before the selected region using the three modifier keys and it will be copied and moved into position immediately before the previously selected region—as shown in this screenshot.

Repeat Region

If you want to create a number of duplicates and place these one after another in the Edit window, you can use the Repeat command. This works similarly to the Duplicate command, but opens a dialog window to let you specify the number of repeats when you choose "Repeat..." from the Edit menu or press Option-R (Mac) or Alt-R (Windows).

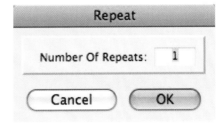

Figure 6.54 The Repeat dialog lets you specify the number of repeats.

Locking Regions

When you have completed your region edits, especially if these include detailed or difficult edits that have taken you some time to create, it can be useful to "lock" the regions or region groups containing these edits to make sure that you don't change anything by mistake. There are two different commands available in Pro Tools that can be used for locking regions: Edit Lock and Time Lock.

Time locking regions is particularly useful when you are using Shuffle mode. In Shuffle mode, Time (or Edit) locked regions, and all regions occurring after the locked region, will stay exactly where they are, even though neighboring regions are being shuffled around.

On tick-based Elastic Audio-enabled tracks, both Time- and Edit-locked regions conform to tempo changes, but other Elastic Audio processing (such as Quantize or manual warping) cannot be applied.

Edit Lock Region

If you want to prevent a selected region or regions, even on different tracks, from being edited (cut, deleted, separated, or trimmed) by mistake, you can use the Region menu's Edit Lock command to prevent this. Edit-locking regions also prevents these regions from being moved to different time locations.

You can copy and paste Edit-locked regions to another track or time location, but the copy will also be Edit-locked at this new time location.

To Edit lock (or unlock) a region, use the Time Grabber to select the region and choose Edit Lock/Unlock from the Region menu, or press Command-L (Mac) or Control-L (Windows). A small Edit Lock icon is displayed in the lower left corner of the region to indicate that it cannot be moved, deleted, or edited.

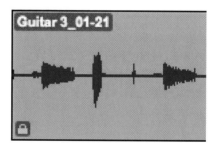

Figure 6.55 An Edit-locked region displays a small lock icon.

Time Lock Region

If you simply want to lock a region to a particular location in a track to prevent it from being moved accidentally, there is a separate Time Lock command for this.

Time-locked regions can be deleted or trimmed, AudioSuite processed, or even moved to the same time location on another track. If you separate a Time-locked region, the new regions that you create by doing this will also be locked to their time locations.

To Time lock (or unlock) a region, use the Time Grabber to select one or more regions, which can be on different tracks, and choose Time Lock/Unlock from the Region menu or press Shift-T. The Time Lock icon, which takes the form of an outline of a small lock, is displayed in the lower left corner of the region to indicate that the region cannot be moved from its time location.

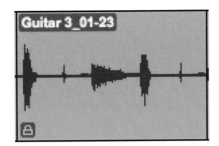

Figure 6.56
A Time-locked region displays the outline of a small lock.

Mute Region

If you want to mute a region, the Edit menu has a Mute/Unmute Regions command that mutes selected regions or un-mutes these if they are already muted. The keyboard command for this is Command-M (Mac) or Control-M (Windows).

Quantize to Grid

Very often during editing, you will find it useful to be able to quantize selected audio or MIDI regions to the Grid. The Region menu's "Quantize to Grid" command is the quickest way to do this. This command adjusts the placement of selected audio and MIDI regions so that their start points (or their sync points, if they contain one) line up exactly with the nearest Grid line. To use this command, choose a Grid value first and use the Selector or Grabber tool to select the regions you want to quantize, making sure that the whole of each region is selected. When you choose the "Quantize To Grid" command, selected regions (or their sync points) are moved to the nearest Grid line. The keyboard command for this is Command-0 (zero) for the Mac and Control-0 (zero) for Windows.

Grouping Regions

Pro Tools lets you group together any combination of audio and MIDI regions on tracks that are next to each other so that they can be treated as a single group for ease of editing. If you use Beat Detective or the "Separate Region At Transients" command to split your drum tracks into individual "hits," you are going to have lots of small regions to deal with. The same thing can happen with imported REX or ACID files, or with looped takes, when you have been trying to perfect a vocal or a guitar solo or whatever. You can use region groups to gather useful sets of regions together so that you can move them around while working on your arrangement. Region groups are also useful for grouping parts and sections together so that they can be easily copied and moved. So, for example, you might group all the backing vocals together in the first chorus so that you can copy and paste these into subsequent choruses.

> **Note**
>
> A region group is a collection of any combination of audio and MIDI regions that looks and acts like a single region. Region groups can be created on single or multiple adjacent audio, MIDI, and Instrument tracks and can include both tick-based and sample-based tracks.

To create a region group, select one or more regions on one or more tracks and choose the Group command from the Region menu.

Figure 6.57
Using the Group command to group regions.

The region group will appear as one region with the Region Group's icon in the lower left corner and will be added to the Region List.

Figure 6.58 A Grouped region selected in the Edit window, identifiable by its Region Group icon in the lower left corner.

You can "nest" region groups together with other regions or region groups. Take a look at the screenshot below and observe the region group created in the example above about to be nested within a new region group comprising this region group and three more regions.

Figure 6.59 Nesting a region group together with other regions into a new region group.

When you choose the Group command, all three groups are joined together to form a new region group.

Figure 6.60 A new Region Group created from an existing Region Group together with two more regions.

The Region menu also has commands to let you ungroup and regroup regions. If you apply the "Ungroup" command to nested region groups, it will only ungroup the top-layer region group, leaving any underlying region groups untouched. The "Ungroup All" command, on the other hand, will ungroup a region group together with all of its nested region groups.

Figure 6.61 Using the Ungroup All Region command.

As you can see from the screenshot below, after using the "Ungroup All" command, the grouped region is disassembled so that the original regions are accessible again.

Figure 6.62 Regions all ungrouped.

With the regions ungrouped, you can edit these individually, as necessary. As long as you don't group and ungroup any other regions beforehand, you can then use the Regroup command to regroup these as they were and continue working on your arrangement.

You can also create multitrack region groups by selecting regions on multiple adjacent tracks.

Figure 6.63 Preparing to create a mixed multitrack Region Group.

After selecting the regions, choose the Group command from the Region menu or press Command-Option-G (Mac) or Control-Option-G (Windows).

Figure 6.64
A multi-track region group created by grouping regions across two tracks.

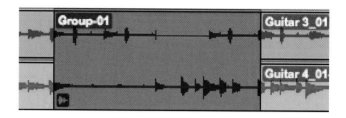

If you group a mix of sample- and tick-based audio or MIDI tracks together, a different region group icon in the bottom left corner of the region group indicates this.

Figure 6.65
Selecting a sample-based audio track and a tick-based MIDI track.

As you will see from the accompanying screenshots, when two mixed regions are grouped like this, a special mixed Group icon is displayed in the lower left-hand corner of the grouped region.

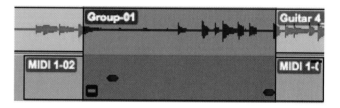

Figure 6.66 A Mixed multi-track Region Group containing both sample-based and tick-based tracks with a mixed region group icon in the lower left corner.

> ## Note
> Region groups can become separated if you move a track from the group so that it is no longer adjacent, hide a track from the group, insert a track within the group, delete a track from the group, record into a region group, change the tempo of a mixed group, or change playlists on a track within the group. A broken icon appears in the lower left corner to warn you if this has happened.

In the example here, the audio track is sample-based, whereas the MIDI track is tick-based. When I changed the tempo after grouping these, the tick-based MIDI track moved along the timeline, while the sample-based audio region stayed at the same location on its track. The "broken" icon appeared in the lower left corner of both regions to warn that this had happened.

Figure 6.67
A mixed multitrack Region group separated by changing tempo.

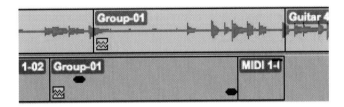

> ## Tip
>
> Pro Tools can export and import region groups using the new ".rgrp" region group file format. This can be very useful for creating multitrack loops containing references to all the audio files within the region group, region names and relative location in tracks, track names, fades and crossfades, and all the MIDI data from the region group. You can then import these into your current or future projects by dragging and dropping the region group file from a DigiBase browser or from Windows Explorer or Macintosh Finder to the Timeline, a track, the Track List, or the Region List.

> ## Note
>
> Region group files don't store automation, plug-ins, track routing, tempo or meter maps, or Region List information.

Dynamic Transport Mode

Dynamic Transport mode lets you decouple the playback location from the Timeline selection. It is similar in this respect to using the "Unlink Timeline from Edit Selection" feature, which is very useful for MIDI editing. But Dynamic Transport mode offers a lot more than unlinking these selections: the timeline selection can be freely moved around by clicking and dragging the gray area that defines the selection in the Main Timebase ruler, for example, and you can start playback from anywhere on the Timeline without losing your Timeline or Edit selections.

So, you can cycle around a section of your song in Dynamic Transport mode while you do some edits on-the-fly, then simply grab and drag the timeline selection to another part of the song to do more edits there. The Dynamic

Transport has a separate Play Start Marker that you can move to a different location on the Timeline—inside or outside of the Timeline selection. If you move the Play Start to within the Timeline selection that you are cycling around, then start playback, it starts from this location, plays to the end of the Timeline selection, then subsequently loops around the whole of the Timeline selection.

A more practical way to use this is to move the Play Start to a bar or two before the timeline selection so that it acts as a preroll to the cycle around the Timeline selection. So if you want to record, say, a MIDI part in 4 bar sections, each with a 1- or 2-bar preroll, you can set the Timeline selection to cover 4 bars, set the Play Start Marker a bar or two before this, record this section, then grab and drag the Timeline selection to the next 4 bars, record this section, and so on. If you want to change the length of the Timeline selection, just drag the start or end markers for the selection back or forth along the Timeline so that it encompasses however many bars you want.

> ## Note
>
> When you enable Dynamic Transport mode, it disables the Link Timeline to Selection function. This allows you to play back either the Timeline selection or an Edit selection—whichever you prefer. You can press the left bracket key to audition the Edit selection and use the Enter key to play the timeline selection. This lets you carry on making your edits, very flexibly, while Pro Tools is playing back. And if you want the Timeline selection to follow the Edit selection, you can always re-enable the Link Timeline to Selection button whenever you like. Then the Dynamic Transport will follow each time you make a new selection in the Edit window.

Using Dynamic Transport Mode

You can enable Dynamic Transport mode by selecting this from the Options menu. If you prefer to use a keyboard command, you can press Command-Control-P (Mac) or Control-Start-P (Windows) to toggle Dynamic Transport mode on or off. When Dynamic Transport mode is enabled, the Main Timebase ruler expands to double-height and reveals the Play Start Marker, which marks the location from which playback will begin when you start playback.

How the Play Start Marker Works

The Play Start Marker is very useful while you are playing back Edit and Timeline selections, allowing you to commence playback independently of these selections. For example, the accompanying screenshot shows the Play Start Marker located approximately halfway between bars 6 and 7 in this example, which has a time signature of 12/8. The Edit selection, as shown by the darkened region in the Edit window and by the Edit Selection Start and End

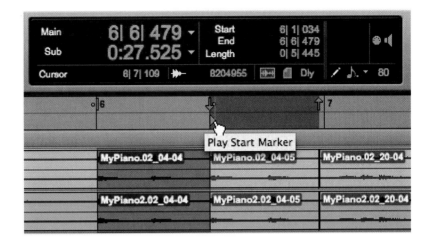

Figure 6.68
When Dynamic Transport Is enabled, the Main Timebase ruler expands to reveal the Play Start Marker.

locations displayed in the Toolbar, does not play back when you press Play: the Timeline Selection, shown by the darkened area in the Main Timebase Ruler (the Bars|Beats ruler in this example) plays back instead.

You can position the Play Start Marker anywhere you like—it doesn't have to be at the start of the Timeline Selection or at the start of the Edit Selection. You can even reposition the Play Start Marker during playback and have playback continue from the new location.

In the accompanying screenshot, playback will start from the Play Start Marker, play through the Edit Selection that follows this, play through the Timeline Selection that follows this, and loop around the Timeline Selection until you stop playback.

Figure 6.69
Play Start Marker positioned independently of the Edit Selection and of the Timeline Selection.

If you want to judge how well the transition into the looped Timeline selection works, but you don't need to start from quite so far back next time, you can simply drag the Play Start Marker, during playback, to a new location and when you let go of the mouse, playback will jump to that location and continue playback from there.

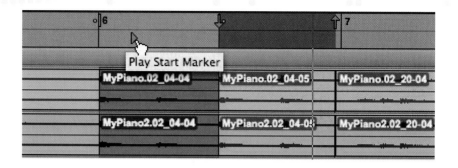

Figure 6.70
Dragging the Play Start Marker to a new location: this even works during playback.

> ## Tip
> To change the Timeline selection, you can either click and drag a new selection in the Main Timebase ruler using the Selector tool, or, with any Edit Tool selected, just click and drag the Timeline Selection Start or End Markers.

To reposition the Play Start Marker, you can click anywhere in the Play Start Marker Strip in the Main Timebase ruler, or, with any Edit Tool selected, you can simply drag the Play Start Marker to wherever you like.

> ## Note
> The Play Start Marker Strip occupies the lower half of the expanded Main Timebase Ruler in Dynamic Transport Mode.

Keyboard Shortcuts for Dynamic Transport Mode

Because it is very probable that you will want to locate the Play Start Marker to the beginning or end of the Timeline or Edit selections, either during playback or when the Transport is stopped, there are several keyboard shortcuts provided for these and similar actions.

Pressing Period (.) on the numeric keypad then the Left Arrow moves the Play Start Marker to the Start of the Timeline Selection: pressing the Right Arrow moves it to the End of the Timeline Selection instead.

Pressing Period (.) on the numeric keypad then the Down Arrow moves the Play Start Marker to the Start of the Edit Selection.

You can nudge the Play Start Marker backwards along the Timeline, a bar at a time, by pressing 1 on the numeric keypad. To go forwards, press 2.

Assuming that you have Bars|Beats selected as the Main Timebase ruler, you can move the Play Start to a specific bar using the numeric keypad: hold the Asterisk (*) key while you type the bar number, then let go of these keys and press the Enter key.

Temporarily Linking the Timeline and Edit Selections

Whenever you have an Edit Tool selected, you can move the Timeline selection backwards or forwards along the Timeline by clicking and dragging. When the Timeline and Edit selection are not linked, only the Timeline selection will move. If you want to temporarily link the Timeline and Edit selections while you move the Timeline selection, press the Option key (Mac) or Alt key (Windows) while you drag the Timeline selection. The Edit selection will be made to correspond with the Timeline selection, and they will move together to the new location.

Recording in Dynamic Transport Mode

If you are recording in Dynamic Transport mode, you can start independently of the Timeline selection, using the Play Start Marker as a manual preroll before the Timeline selection, and Pro Tools will punch in to record at the start of the selection and punch out at the end of the selection.

Setting Selection Start and End in Dynamic Transport Mode

When you enable Dynamic Transport mode, Pro Tools automatically enables Loop Playback mode and automatically disables "Link Timeline and Edit Selection." As a consequence, when you select a region in the Edit window, the Start and End points are *not* automatically entered into the Transport window's Play Selection fields or the Toolbar display's Edit Selection Start and End locations. However, a special keyboard command becomes available in Dynamic Transport mode to let you do this manually: just press the letter "O" on your computer keyboard.

Auditioning Loops in Dynamic Transport Mode

You can audition a series of loops without stopping and restarting playback when Dynamic Transport is enabled. Simply select a new region that you wish to loop around during playback and press the letter "O" to update the loop start and end locations into the Transport window and Pro Tools will immediately jump to the start of the new loop region.

Preferences for Dynamic Transport

If you want the Play Start Marker to always snap to the Timeline Selection Start Marker when you move the Timeline Selection, or when you draw a new selection or adjust the Timeline Selection Start, you can set a preference for this— "Play Start Marker Follows Timeline Selection"—in the Operation Preferences window. The way this works when you are playing back is that if you click anywhere in the Edit window, playback will immediately jump to that position—great for quickly checking through the session to see how everything is sounding while comparing sections. Similarly, if you drag in the Edit window to make a selection, playback will immediately jump to the beginning of that Edit selection.

Figure 6.71 Selecting "Play Start Marker Follows Timeline Selection" in the Operation Preferences.

Another important setting to choose in the Operation Preferences window is the playback behavior, that is, what happens when you stop playback. When the option "Timeline Insertion/Play Start Marker Follows Playback" is selected in the Operation Preferences window, the Timeline Insertion and the Play Start Marker both move to the point in the Timeline at which playback is stopped. So when you start playback again, playback continues from the point at which you stopped.

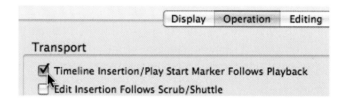

Figure 6.72 Selecting "Timeline Insertion/Play Start Marker Follows Playback" in the Operation Preferences.

This is such a regularly used preference that a dedicated button is available to let you toggle this on and off from the Edit Window toolbar.

Figure 6.73 Setting "Timeline Insertion/Play Start Marker Follows Playback" using the button in the Edit window Toolbar.

If you want the Timeline Insertion and the Play Start Marker to stay at the location from which you originally commenced playback, you need to make sure that this option is deselected. Then, when you start playback again, it starts from this original starting point—not from the place where you stopped.

> **Tip**
> You can toggle the Timeline Insertion/Play Start Marker Follows Playback option on and off using the "n" key on your computer keyboard.

> **Note**
> Be careful not to accidentally hit the "n" key on your computer keyboard or you will find that your session is behaving differently from the way you expect!

Track Compositing

Very often when you are recording or overdubbing, you will record multiple "takes" until you have captured the best possible performance. Maybe this will be the last "take" that you record—and maybe it will be the first! I have often seen producers or musicians insist on repeating takes in their search for the perfect performance, only to discover with hindsight that the first take was the best. On the other hand, I have witnessed just as many recordings in which the musicians only achieved their best performance after many attempts. And with others, it could be the middle take that is best, starting off with takes that are not perfect, reaching the best performance achievable, then tailing off after this point.

In many, many cases, the musicians will mess up one section but perform brilliantly in another section. This is when you will need to combine material from two or more takes to make a composite of the best material that you have recorded. Typically referred to as "comping" the takes, a more formal way of referring to this would be "track compositing."

You may have recorded the various takes into different tracks, but it is more likely that you will have recorded these into different playlists on the same track or tracks. You may also have used Loop Record to record multiple takes into the same section of the music, perhaps for a guitar or sax solo.

> **Note**
> A "playlist" contains all the regions in a track in the locations at which they were recorded or to which they have been moved.

If you will be using Loop Record and want to be able to use the track compositing features, you should enable the Operation Preference to "Automatically Create New Playlists When Loop Recording." Before you start recording, enable Loop Record mode and select a range over which to loop.

Figure 6.74
Loop recording.

After using Loop Record to automatically record new takes into new playlists, you can switch the track to Playlists view where you will see all the takes in the alternate playlists below the main playlist. You can select the best parts from these alternate playlists and copy them to the main playlist to create the "perfect take."

If the Automatically Create New Playlists When Loop Recording option was disabled before you started loop recording, select the main playlist and Command-click (Mac) or Right-click (Windows) the region in the main playlist and choose Expand Alternates to New Playlists from the Matches submenu.

Figure 6.75
The Matches
submenu.

The first of these alternate playlists will show a long region containing all of the takes in one long file. The other alternate playlists will contain the individual takes with their start times set to match that of the Edit selection around, which you were looping when you recorded these.

Figure 6.76 The Alternate Playlists revealed.

You don't really need to see the whole file region that contains all the takes—you just want to see the individual takes that have the same start and end times as the Edit selection. To hide this long region, you can Control-click (Mac) or Right-click (Windows) on the Playlist name and choose Hide.

Another way to do this is to select from the main playlist the same Edit selection that you used for loop recording, then Control-click (Mac) or Right-click (Windows) the Track Name or any of the Playlist Lane Names and choose "Show Only Lanes With Regions Within the Edit Selection" from the Filter Lanes submenu.

Choosing the Takes

When you have your takes showing in the alternate playlists in the "Comp lanes" below the main playlist, you will probably want to audition these in Loop Playback mode, checking each take in turn by pressing the button marked "S" on each Comp lane. The "S" stands for "Solo," but pressing this does not solo the alternate take, it mutes the main playlist and plays the audio on the Comp lane instead—and you also hear the rest of your mix at the same time.

You can press Shift-S to solo any Comp lane containing the Edit cursor and you can use Control-P and Control-";" (Mac) or Start-P and Start-";" (Windows) to move the Edit cursor or the Edit selection up and down through the Comp lanes.

Figure 6.77
Copy Selection
To Main Playlist
button.

When you hear a section on a Comp Lane that you prefer, you can select this and click the upwards-pointing arrow to the right of the Solo button to copy this selection to the main playlist.

Figure 6.78
Selection moved
to main playlist.

Playlists View

Each track in the Edit window has a Playlists view that shows the main playlist. Any alternate playlists associated with the track are displayed directly below the track in Playlist lanes.

These alternate playlists can be edited in the same ways that you can edit the main playlist in Waveform view. When you apply edits to range selections in Playlists view, these are applied to all alternate playlists that are shown. Alternate playlists that are not shown are not affected.

Tip

Before starting work with Track Compositing, duplicate the track's Main Playlist to keep it intact as a backup alternate playlist.

To view the Playlist lanes for a track, select Playlists from the Track View selector.

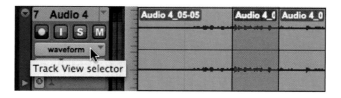

You can also Command-Control-click (Mac) or Control-Alt-click (Windows) the Playlist selector to switch to Playlists view.

If a track contains one or more alternate playlists in addition to the main playlist, these will be displayed in lanes underneath the main playlist. The bottom Playlist lane is always empty and can be used for adding regions to create new alternate playlists.

> **Note**
>
> If you have not created any alternate playlists for a track, you will still always see one empty playlist when you switch to Playlists view.

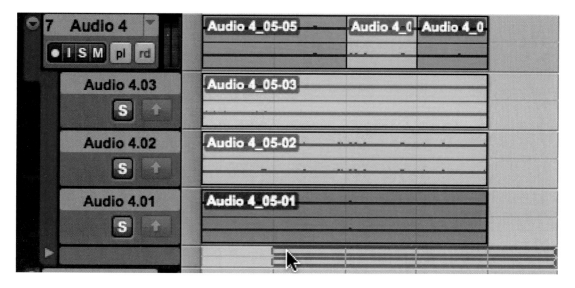

Figure 6.81 Adding another region to the bottom playlist.

Copy Selection Commands

Once you have found a selection in an alternate playlist that you want to use in the main playlist, you can always manually copy and paste the selection to the main playlist to construct the best combination of selections from the different takes. Pro Tools also provides three Edit menu commands that you can use to copy and paste the selection.

Figure 6.82
"Copy Selection To...." commands in the Edit menu.

The "Copy Selection to Main Playlist" command copies and pastes the selection to the main playlist, overwriting any material at that location. You can use this menu command as an alternative to using the buttons in each Comp lane.

The second command, "Copy Selection to New Playlist," creates a new, empty main playlist and copies and pastes the selection to the new main playlist. A dialog box opens when you use this command to let you type a name for the new playlist. I often use this command when I find the first section that I want to use and name this as my "Comp" track. Then I go through all the takes and just add the sections that I want to use to this playlist.

Figure 6.83
A new playlist created using the "Copy Selection to New Playlist" command.

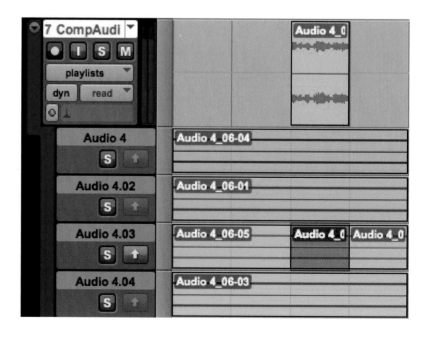

The "Copy Selection to Duplicate Playlist" command works in a similar way, but in this case, it duplicates the current main playlist and copies and pastes the selection to the duplicate main playlist. As with the "Copy Selection to New Playlist" command, the previous main playlist moves to a new Playlist lane.

Why You May Need to Filter the Comp Lanes

Compositing tracks using these features can result in you having large numbers of playlists in the tracks you are working with. Say you loop record the vocalist on the verses and do four takes, then you do four more on the bridge sections, then four more on the chorus sections. Already that is 12 takes, and when you are comping the verse sections you probably don't want to view the bridges and choruses at the same time. This is where you will value the filtering feature.

You can filter Playlist lanes to show or hide them based on different criteria such as the Region rating or whether they are inside or outside of the Edit selection. To filter Playlist lanes, you can either Control-click (Mac) or Right-click (Windows) the Track Name or the Playlist Name to access the context menu. If you access the menu from the Playlist Name, you get options to hide, scroll into view, create, delete, or rename the playlist, or to Filter the Lanes.

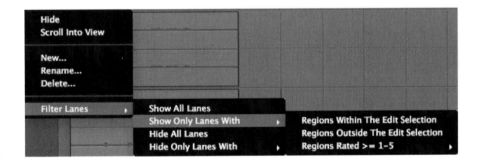

Figure 6.84 Control-clicking the Playlist name opens this menu.

If you are going to filter based on ratings, it helps to view the ratings in the regions, and you can select this option in the Regions submenu from the View menu. You can choose a rating, from 1 to 5, for any region by selecting the region and choosing a rating from the Region menu or by pressing Command-Control-Option (Mac) or Control-Alt-Start (Windows) and then typing a number between 1 and 5.

Figure 6.85 Viewing a rating in a region.

Compacting Audio Files

When you have finished your editing session you should consider "compacting" any edited audio files to remove any unused audio that you are sure you have no further use for. This makes the files smaller, so it saves disk space from being used up unnecessarily.

If you have recorded or imported a large number of audio files, yet have only actually used a fraction of these, compacting the files can save considerable amounts of disk space, making backups much less costly both in time needed to make the backups and for the media used.

To compact an audio file, select the region or regions that you want to compact in the Region List and choose Compact from the pop-up menu at the top right of the Region List.

In the dialog window that appears, you can *pad* the regions within the compacted file to include some of the original audio (if this exists) at either side of the boundaries of the selected regions. You may have placed the region boundaries too close to the edit points—cutting off very quiet sounds at the beginning or end of the regions. By including some of the audio just before and just after the region boundaries, you ensure that this audio exists in the new file that you are creating—and you can always make more careful edits to remove this extra audio at any time in the future. So, it is a wise thing to enter an amount of padding, in milliseconds, that you want to leave around each region in the file.

When you click Compact, the file or files are compacted and the session is then automatically saved.

> **Note**
>
> Because it permanently deletes audio data, the Compact Selected command should only be used after you have completely finished your editing and are sure that you have no further use for the unused audio data.

Summary

The best way to learn how to edit in Pro Tools is, of course, to edit in Pro Tools. There is just no substitute for hands-on experience here. The best way to use this chapter is when you are sitting at your computer with Pro Tools right there in front of you so you can try each editing move that I have described here for yourself.

Obviously, you are not going to learn everything in this chapter overnight, and you won't need to use all these editing techniques with every session that you work on. But if you go through this chapter thoroughly, trying as many of the editing commands and techniques as possible for yourself, there is a good chance that you will remember that a particular command or technique exists when the need for this arises.

You can also use this book as a handbook to keep around whenever you are using Pro Tools so that you can dip in and remind yourself of how a particular feature works when you need to use this.

In This Chapter

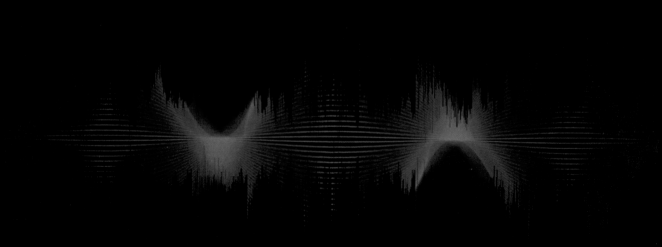

7

Mixing

Introduction

There are many approaches to mixing music that are recommended in books and by respected pundits. This chapter will not provide prescriptions for how to mix your music. Instead, the first section reveals various technical issues that you should consider while preparing for your mix. This is followed by the discussion about the various "tools of the trade" that you may use during your mixing sessions. The third section gives suggestions about how you might progress while you are organizing your mix, marking out the sections, ordering the tracks, grouping tracks together, setting up VCA groups, and using the Edit window to automate the mix. Finally, you get to the mix itself. You may be using a hardware controller together with the real-time Mix Automation features, or you may mix "manually" by setting the faders by hand, or by drawing in automation curves with the mouse. Ultimately, you will arrive at a result that you are happy with. At this point, you will make your "final" or "master" mix—which is the one that you will supply to the Mastering Studio that prepares the mixes for release on CD and in other formats. If you are intending to "master" the mixes yourself, you may still choose to make these "final" mixes at the end of the music production process and handle "mastering" as a separate stage once the production is finished.

Mixing is usually considered to be the final stage in the music production process—the end of the journey in many cases, but not always. In some genres of music, remixes are made for a variety of reasons. The most basic reason would be that the producer, artist, or record label feels that the first mix was simply not good enough. Another reason would be to provide the audiences with alternative versions to listen to, when the music is broadcast on the radio or played by a club DJ. It is also quite common for a producer to hear the music, possibly years after it was first released commercially, and be enthused by the idea of creating a new mix—sometimes even a radically different mix that uses only the vocal and replaces all other music elements.

Preparing for Your Mix Session

Before you get started, you need to make sure that your monitors are set up correctly, consider what monitor levels to mix at, understand how your meters work, and line up the meters on any connected analogue equipment. To check mono compatibility, you need to understand how to use a Phase Scope, and if you want to avoid distortion problems, which can arise with "hot" mixes producing a "fatiguing" sound that can be very unpleasant for the listener, then you need to know how to use an oversampling meter. If you are not using an external mixer, you will need a monitor controller to provide a convenient volume control for your playback listening level; and if you want to add an analogue "sound signature" to your mix, one way to go is to use a summing mixer.

Check Your Monitoring

One of the most important things to attend to is your monitoring system—you have to be able to hear properly what you are doing when you are mixing. This means, you have to make sure that you are sitting in the "sweet spot" at one of the three corners of an equilateral triangle with the left and right speakers positioned at the other two corners facing you, and with the tweeters at about the same height as your ears.

You should also make sure that the amplifiers have more than enough power to reproduce sound at the highest listening levels you intend to use without significant distortion and that the speakers have a response as "flat" as possible over a range as wide as possible. The so-called "nearfield" monitors (such as the previously-ubiquitous Yamaha NS10s) that sit on stands, on desktops, or on the mixing console's meter bridge can never reproduce the full-range of the audio spectrum—because they have to be relatively small. This means that they will not be able to reproduce the bottom octave properly, as their response will roll off significantly below, say, 80 or 100 Hz. This is why professional studios have large, full-range monitors to make sure that the engineers and producers can hear all the frequencies correctly.

You also need to take into account the acoustics of the room that you are listening in. For example, consider the situation where you are trying to set the level of a bass drum correctly for your mix. The fundamental frequency of this instrument might be around 80 Hz, or even lower. The length of the sound wave that is produced at this frequency (the wavelength corresponding to this frequency) would be somewhere around 4 m, which could easily correspond to the width of the room that you are listening in. A "standing wave" would very likely exist between the walls of your listening room at this frequency. Consequently, if you are sitting at a point where the amplitude of this wave is zero or very low, the bass drum will sound too quiet, and you will almost certainly compensate by making this too loud in your mix. This is exactly what happens in my home studio, for example. Conversely, if you are sitting at a position within the standing wave where the amplitude of the wave is at a maximum, you will think that the bass drum is too loud—and you are likely to compensate for this by making the level of the bass drum quiet in your mix.

Rooms can be designed or modified to minimize these effects and monitoring systems can be adjusted to compensate for the room acoustics. Professional studios offer properly designed rooms and monitoring systems for hire at professional prices. However, smaller project studios and home studios often overlook, or the owners cannot afford, such necessities. If your listening environment and equipment are compromised, particularly at the low frequencies, then you will have to listen to other systems that will reveal problems at the low end (or elsewhere in the frequency ranges, for that matter).

> **Tip**
>
> I always check my mixes on the stereo system in my car, which instantly reveals whether I have made the bass drum and bass guitar too loud or not when compared with typical commercial albums. Then I go back and adjust the levels on my mixes, make another test CD, and check this in the car until everything sounds correct.

It is always a good idea to listen to your mixes on the kinds of speakers and in the kinds of environments that will ultimately be used by your audience. So, you might take your mixes out to a club, or play them in your car, or try them on your living room system—and get the idea. And don't forget to check your mixes in mono—especially if you want them to sound right on radio, TV, or film where mono playback is still sometimes encountered.

> **Tip**
>
> Find a great mastering studio to check your mixes in: professional mastering studios typically have rooms designed to minimize acoustic problems, with monitoring systems installed costing many thousands of pounds. Many smaller studios now provide "stem" mixes to mastering studios so that the levels of the vocals, bass and drums, and other instruments can be separately balanced in a revealing acoustic environment.

A Note on External Mixing/Monitoring

It is possible to connect a Pro Tools|HD audio interface directly to your monitoring system, but this is not a good way to set your system up because the only way to conveniently control the volume of your monitors would be to lower the Master Fader (or the channel faders).

If you lower the levels in your mix, just to make the sound in the room quieter, then you are almost inevitably going to end up making the level of your mix too low when you save this as a file, route the mix to an external recorder, or whatever.

Also, although there may be volume controls on self-powered monitors or on the power amplifier that is driving your passive monitors, these controls are not normally going to be easy for you to reach while operating the system.

As most engineers will be aware, Pro Tools, like most DAWs, is designed to give best results when the Master Fader, which controls the summed output level, is kept around unity gain (the 0 dB fader position), while the channel faders set the levels of the individual mix elements in relation to this. Although Pro Tools allows you to lower the Master Fader substantially without significant loss of resolution, it is not a good working practice to do so. For example, if the converters are operating at very low levels, any non-linearity will be most pronounced.

One solution is to use some kind of external monitoring control unit that allows you to connect the outputs from your Pro Tools interface and provides a volume control that controls its outputs to your monitors. Such units may allow you to connect and switch between two or more sets of monitors and may provide additional talkback, dimming, or foldback/headphone facilities.

Another solution is to use some form of external mixer that incorporates these facilities. Using an external mixer provides another advantage: it allows you to completely avoid any latency from the Pro Tools system while recording from external sources, because you can route these into the external mixer, monitor them directly from the mixer (with no latency delay), and route them in turn into the Pro Tools to record.

Most professional mixers also incorporate a number of reasonably high-quality microphone preamplifiers. These can be very useful if you are regularly recording from "live" sources—these are also cost-effective compared with the price of buying lots of dedicated microphone preamplifiers.

Choose Your Monitoring Levels

There are no standard monitoring levels followed in music recording studios—unlike in film dubbing theatres, where the Society of Motion Picture and Television Engineers (SMPTE) has established a Sound Pressure Level (SPL) of

85 dB as a standard. This is normally referenced to an electrical signal level of −18 dBFS = 0 VU, which is the standard Operating Level recommended for digital systems by the Audio Engineering Society (AES). For the smaller rooms typically used for mixing music, a Sound Pressure Level of 83 dB may be more appropriate, with a nominal operating level of −20 dBFS providing 20 dB of headroom above this to accommodate signal peaks.

I have been at the sessions in music studios that use monitoring levels well up to the 90s of decibels of SPL—or even more than 100 dB. Even relatively short exposure times to such high levels can make your ears "sing" because of the temporary threshold shift. Nevertheless, you may wish to check out the sound of some loud instruments that you have recorded at somewhere near to their original sound levels. And instruments such as the trumpet or snare drum can easily reach levels of over 100 dB; hence, it can be useful to have a monitoring system that is capable of reproducing such realistic levels.

Many engineers like to mix on small "nearfield" monitors at quite low levels at somewhere between, say 65 and 75 dB SPL. This makes it easier to tell whether the lead vocal or lead instrument can be heard properly at all times and whether other important elements such as the bass guitar and snare drum are at the right levels in relation to the lead. This ensures that the mix sounds great when played back on domestic speakers after the record is released to the public. Of course, you can always check from time-to-time how it sounds at much higher volumes or on larger speakers.

> ## Tip
> I personally mix on "nearfields" around 83 dB or sometimes a little higher— checking this level occasionally using a small hand-held SPL meter. When I have a balance that is sounding good, I hit the dimmer switch on my Yamaha DM1000 mixing console to drop the monitoring level down by 20 dB to 60 or 65 dB and then check that everything I want to hear is still audible.

Line-up Any Connected Analogue Equipment

If you connect your Pro Tools system to any analogue devices, such as mixers, tape recorders, or effects devices, you must make sure that the levels on the meters line up correctly. A 1 kHz test signal sent from Pro Tools at the chosen reference level, such as −20 dB relative to Full Scale, should produce a reading of 0 VU on an analogue VU meter connected to an analogue output from your Pro Tools interface. If it doesn't match exactly, you will need to "tweak" the calibration control on the VU meter until it does match—this is the procedure for "lining up" the meters.

> **Note**
>
> Remember that the meters in Pro Tools read signal peaks—they are not reading an average of the signal like a VU meter does—and there is no "0 VU" nominal operating level marked on the meters in Pro Tools.

To check that the connected equipment is lined up correctly, you can insert the Signal Generator plug-in into one of the Master Fader Inserts, generate a 1 kHz sine wave "test tone" at –20 dB, –18 dB, or whatever reference level is being used, and see whether the meters on the connected equipment match this.

Figure 7.1
The Signal Generator plug-in can be used to generate a test tone.

> **Note**
>
> If you connect your Pro Tools system digitally to external digital devices such as mixers, DAT (Digital Audio Tape) recorders, or effects units and if these use the same digital standards, then the signal levels sent from Pro Tools will automatically produce the same meter readings on these destination devices and vice versa.

Recording a Test Tone to Disk

You can always record a test tone to disk, which can be useful if you transfer recordings to other systems, especially older analogue systems that need to be calibrated with their VU meters lined-up correctly.

To record a test tone to disk, simply insert the Signal Generator on an Auxiliary Input and bus its output to the input of an audio track.

Creating a Test Tone Using AudioSuite

If you prefer, you can use the Signal Generator AudioSuite plug-in to generate a 1 kHz sinewave at a suitable level to use as a reference tone so that other equipment can be aligned to, in order to establish the correct value for 0 VU.

Figure 7.2
Recording a Test
Tone at −20 dBFS
to disk in mono
from a Signal
Generator plug-in
inserted on an
Auxiliary Input to
an Audio track.

Make an empty Edit selection in an audio track that lasts for 30 seconds, or for the time you would like the test signal to last, open the AudioSuite Signal Generator plug-in, choose the settings that you require, such as the level, and then press Process to create a file on disk containing the test signal.

You can supply this file along with your mixes—or record the audio from this in real time onto a tape, if you are transferring to tape.

VU Meters, Operating Levels, and Headroom

The "VU" or "Volume Units" meters used in analogue audio systems display an average measure of the signal level that corresponds reasonably well with the way the human ear perceives the volume of the reproduced audio. They do not reveal the peak levels of the audio signals, which may be 12 or more dBs higher than the average levels. Because these peak levels are typically very short in duration, they are often referred to as "transients," meaning that they pass quickly.

The nominal Operating Level of an analogue recording or mixing system is 0 VU, which is used as the standard input and output reference level for signals entering and leaving the systems. This

Figure 7.3
Using the Signal
Generator
AudioSuite Plug-in
to generate a test
tone.

"nominal" operating level is the maximum average signal level that recording engineers, using analogue equipment, aim for when using VU meters, knowing that there will be "headroom" above this to accommodate transient peak levels. The maximum peak levels that these systems can handle without distortion will vary, but would typically be at least 12 dB higher than 0 VU, as with tape recorders, for example. The difference between the 0 VU level and the maximum peak level is called the "headroom."

This operating level may be set differently in different audio systems:it is set some way below the maximum level that the system can handle, to allow "headroom" for transient level peaks. If the signals passing through the system are much lower than 0 VU, they may get too close to the background noise floor that exists in all audio systems. On the other hand, if average signal levels are much higher than 0 VU, transient peaks within these may attempt to exceed the maximum allowable level in the system, resulting in the waveforms becoming "clipped" and the sound becoming distorted. In practice, you should set levels in such a way that the loudest sounds average around 0 VU, that is, without getting too much lower or too much higher. Also, don't forget that most music will normally contain many quieter sounds that may be 20 dB less (or even lesser) in level than the loudest sounds.Of course, you are not seeing the peak levels on VU meters, especially transient peaks that only occur for a fraction of a second, and peak levels can be much higher than the average levels. To completely avoid distortion when recording audio to analogue tape systems, peak levels should not exceed 12 dB above 0 VU. A characteristic of analogue systems such as tape recorders is that the distortion is mild at first—therefore, relatively high levels can be used before obnoxious distortion is heard. Some recording engineers exploit this situation by recording higher levels onto the tape so that some distortion of the waveforms occurs, making the recordings sound "bigger" or "fatter" than perfectly clean, undistorted recordings.

Audio that is being recorded to a digital system from an analogue system with 0 VU set to correspond −20 dBFS has 20 dB of headroom to accommodate signal peaks. However, in digital systems, if any signals exceed the 0 dB Full Scale level, they will be "clipped" to this level as they pass through the system, badly distorting the signal and making it sound extremely unpleasant. So with digital systems, you should always make sure that you have levels set low enough to avoid clipping—even if you have to use compressors or limiters to allow you to increase average signal levels.

When you are working with digital systems, the nominal operating level for any connected analogue audio equipment is sometimes set at 18 dB below the 0 dB Full Scale (0 dBFS) level that represents the maximum allowable signal level in the digital system. This provides headroom of 18 dB higher than the 0 VU level on the analogue equipment. Another standard that is recommended when you are mixing audio is to set the nominal operating level at −20 dBFS. This allows 20 dB of headroom for peak levels, encouraging the mix engineer to create mixes with greater crest factors (peak-to-average ratios).

Bomb Factory Meter Calibration Simulation

To see how this works, you can use the Bomb Factory Essential Meter Bridge. This simulates an analogue VU Meter and has buttons to let you switch it to emulate different operating Levels.

Just as an exercise, I played back a 0 dBFS, 1 kHz sine wave test tone from a mono file on disk, using an audio track with its fader set to unity gain, through a mono Master Fader which is also set to unity gain. I inserted the Bomb Factory Meter Bridge on the Master Fader and switched this to read the signal peak with no calibration adjustment. As you would expect, this showed a peak level of full scale.

Note that in this mode of operation, the Meter Bridge is not working as a VU meter, which would average the incoming signal over time in some way, thereby producing a reading somewhat lower than the full scale: it is working as a peak meter, just like the Pro Tools meters.

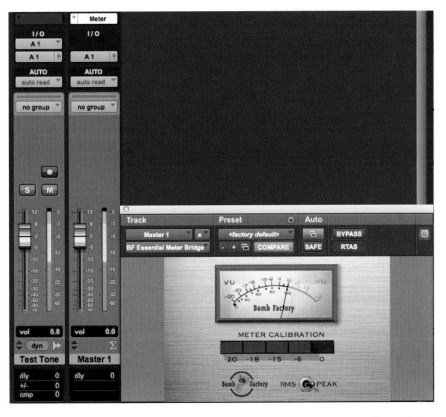

Figure 7.4 Playing back a 0 dBFS, 1 kHz sine wave test tone from a mono file on disk using an Audio track with its fader set to unity gain via a mono Master Fader also set to unity gain with a Bomb Factory Meter Bridge inserted and switched to reading signal peak with no calibration adjustment.

If you connect a real analogue VU meter to an analogue output from your Pro Tools interface, you would need to calibrate this meter to make sure that when you send the agreed reference level from Pro Tools, it would read 0 VU. One such standard reference level is to have 0 VU correspond to –18 dB FS. To see how this will look on an analogue meter, you can use the Bomb Factory Meter Bridge to simulate a VU meter. Send a –18 dBFS reference tone from Pro Tools to the VU meter. On a real VU meter, you would then adjust a potentiometer until the needle lines up exactly with 0 VU. This is called the "line-up" procedure because you are "lining up" the needle with the zero mark.

The Bomb Factory Meter Bridge plug-in allows you to switch between various commonly used meter calibration levels using the row of buttons below the meter. In this example, I am using –18 dBFS. As this is merely a digital simulation, you won't have to tweak anything to get the Bomb Factory meter to read exactly 0 VU—assuming that you are sending it exactly the reference level that you have chosen and that you have put the meter calibration switch in the corresponding position. With a real VU meter, you would almost certainly need to tweak the level showing on the meter at least slightly to get it to line up exactly with the 0 VU mark.

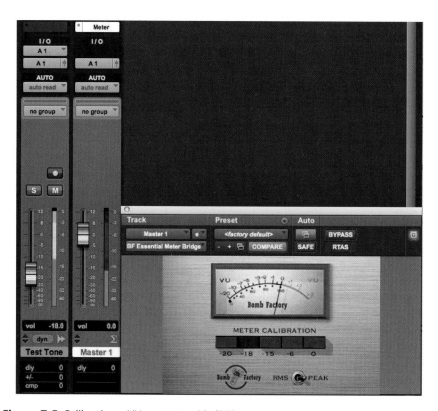

Figure 7.5 Calibrating a VU meter to –18 dBFS.

Bomb Factory VU Meter Simulation

To use the Bomb Factory plug-in to meter an audio track or a mix in Pro Tools in VU mode, you need to switch it from Peak to RMS (Root Mean Square)—which shows an average value for the signal—and make sure that you have the Meter Calibration button selected to correspond to the reference level that you would be using with external analogue equipment, such as −18 dB FS. Then you will see the needle fall back slower than the peak meters in Pro Tools and it will vary above and below the 0 VU mark—making use of the 18 dB of available headroom above 0 VU. The average level of your mix should be around the 0 VU mark, and excursions above this will not normally be a problem because 0 VU represents −18 dB FS, leaving you a further 18 dB of headroom above 0 VU to accommodate excursions and transient peaks.

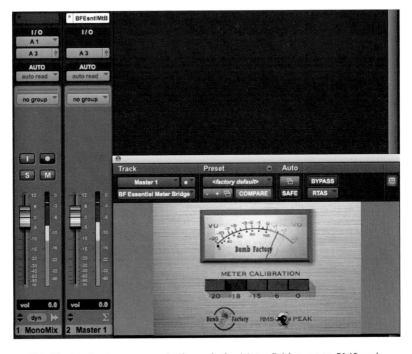

Figure 7.6 Playing back a mono mix through the Meter Bridge set to RMS and calibrated to −18 dBFS reveals how a VU Meter would behave—showing the metering varying above (it also varies below) the 0 VU mark in response to the Audio in the mix.

Metering in Pro Tools

Each Pro Tools track has its own metering running vertically next to the track's fader. These meters let you see the signal levels on individual faders, or what the summed signal levels are at the Master Faders.

Note

The difference between the nominal operating level and the level of background noise inherent in the system (often called the "noise floor") is referred to as the "signal-to-noise ratio." Typically, digital systems have a much lower noise floor than analogue systems. As a consequence, you are more likely to hear quiet sounds—such as rattles or squeaks from musical instruments or drum kits, or breathing, or other sounds made by musicians—that you may not have noticed with analogue recordings where these low-level sounds may have been masked by the general background noise, tape hiss, or other noise signals.

Note

Theoretically, the dynamic range of a 16-bit digital system is 96 dB while that of a 24-bit system is 144 dB. The dynamic range is the range between the minimum and maximum signal levels that a system can handle.

Audio tracks default to pre-fader metering, so the meters on audio tracks show the levels of signals being presented at the track inputs from the audio interface or being played back from audio files on disk.

On Auxiliary Inputs, Instrument tracks and Master Faders, the meters indicate the level of the signal being played through the channel output or outputs.

Signal present is indicated by a green color that turns to yellow from −12 to −3 dB and turns to orange when the level reaches 3 dB or less below full scale. Each meter also has a Clip indicator at the top that lights up red if clipping occurs.

The "dB" markings to the right of each *meter* are peak levels that show 0 dB Full Scale at the top of the range.

The "dB" markings to the right of each *fader* show a unity gain position (no increase or decrease in level) marked "0" (Zero). The fader controls the gain for the channel and this can be increased by up to 12 dB or decreased by 90 dB or more.

Note

A fader controls the gain, or amplification, applied to an audio signal. Increasing the gain above the unity gain level boosts, or amplifies, the signal. Decreasing the gain below the unity gain level attenuates the signal, reducing its level. The amount of positive or negative gain applied to a signal is simply expressed as a ratio in decibels of the output level compared with the input level.

If you set the fader at unity gain on an audio track (zero on the scale), the playback level of the audio coming from the disk is not altered. You can increase this playback level by up to 12 dB or you can decrease it infinitely until you hear no audio playing through the channel—however, you will still see the level being presented to the audio track's input from the audio file on disk showing up, unaltered, in the Track meter.

> **Note**
>
> You can change the metering for all the audio tracks in your session to post-fader metering by deselecting the Pre-Fader Metering option in the Options menu. In this case, the meters show the output level from the track, as controlled by the fader. Thus, if you pull the fader down, the level displayed on the meter will go down, accordingly. This is a less useful option because it does not reveal any clipping that might be present at the input to the track.

If you have audio generated by a plug-in inserted into an instrument or auxiliary track, or coming into an Auxiliary channel from an external source, the fader controls the gain for this in the same way. The meter shows the level being presented to the audio input of the instrument or auxiliary track from the external source or from the plug-in, and the fader controls how much of this signal is passed through to the track output (which usually feeds your mix bus, but could be routed wherever else you choose). So, if you bring down the level of the track fader, the signal level in the Track meter does not alter. But if this track is routed to a Master Fader, for example, you would notice the level in its meter fall, as you reduce the gain on either the track fader or the Master Fader.

> **Note**
>
> Nothing that you do with an audio fader (that only controls the level of signals passed to the output of the track always) will affect the audio in a file on disk playing back via this track. So, if the audio that was recorded into a file on disk is clipped, for example, reducing the playback level will not prevent it from clipping—it will just play back the clipped audio at a lower level—and if the level of audio in the file on disk is sufficiently high, you will continue to see the clip indicator light up in the meter on the audio track.

PhaseScope Metering

In the accompanying screenshot, the DigiRack Signal Generator is shown producing a 1 kHz sine tone at 0 dB FS inserted on a stereo auxiliary input with the fader set at unity gain. The Phase Scope shows a vertical line and the Phase correlation meter shows +1, indicating that the audio in both left and right channels is identical.

The level shown both on the level meters in the Phase Scope plug-in inserted on the Master Fader and on the Master Fader meters has been reduced by 6 dB by lowering the Master Fader—just to demonstrate how this works.

If you look at the meter on the Master Fader, you will see that this has dropped by 6 dB compared with the 0 dBFS input coming from the Signal Generator. You can see this by reading the markings to the right of the meter. If you look at the dB markings to the right of Master Fader, you will see that this has been moved down from the unity gain 0 dB position to the −6 dB position.

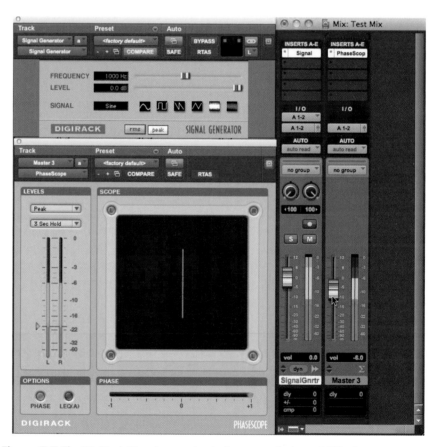

Figure 7.7 The DigiRack Signal generator is shown producing a 1 kHz sine tone at 0 dB Full Scale inserted on an Auxiliary Input with the fader set at unity gain. The level presented to outputs 1-2 via the Master Fader has been reduced by 6 dB by lowering the Master Fader.

Fat Meters

When you are mixing, you may prefer to use the "fat" meters instead of the normal "thin" meters in the Mix (and Edit) windows. To switch between the normal and "fat" meters, hold the Command (Start), Option (Alt), and Control keys and click on any of the meters.

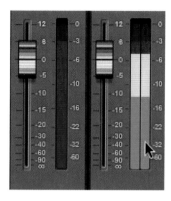

Figure 7.8
"Fat" Meters.

Metering and Loudness Discussion

To help you get a better understanding of some of the issues involved, here are some extracts from an article that I wrote for Pro Sound News Europe after discussions with Thomas Lund at TC Electronic, and mastering engineer Bob Katz in the US:

Thomas Lund has this advice for mixing engineers: "If you apply the old-school sample peak detection methods of measuring level, peaks need to stay below −3 dBFS when delivery is linear, i.e., via CD, and below −5 dBFS when delivery is via MP3 at 128 Kbps. That's not even always enough headroom—but 95% of the time it would be." The message is that, if you mix to digital, don't peak higher than −3 dBFS.

Lund also says "Engineers should keep in mind that highly processed mixes with certain higher frequencies and high peak levels or clipping are likely to lead to levels in excess of 0 dBFS when reproduced at the outputs from typical D/A converters, which can lead to further distortion if there is insufficient headroom in the converters or whenever the signal crosses domains, such as in conversion to mp3." Inter-sample peak metering to view and protect from these types of peaks is now available from TC Electronic in their System 6000 and from Sonnox and PSP in their peak limiters.

New tools are undoubtedly needed to help prevent the hyper-compression (squashed, loud mixes, and masters) arising as a consequence of super-high average levels and low peak-to-average ratios. Mastering "guru" Bob Katz has plenty to say on this subject: "The invention of digital audio started the accelerated loudness race which raised average levels on commercial CD releases almost 20 dB in 20 years. This was caused by the new ability in digital audio to normalize to the peak level. We need to return to the concept of headroom and standardize on an entirely new type of meter that is calibrated to true loudness and which allows for adequate headroom, with the true peak level hidden from the user."

Katz, here, is referring to the situation that peak-only metering—especially in software DAWs such as Pro Tools, certain CD players, and other digital equipment—is no longer "fit-for-purpose." Katz's proposed alternative, the K-System, incorporates RMS metering and is coordinated to a calibrated monitor gain. RMS metering is more accurate than simple averaging, although not as accurate as a true loudness meter—which requires more DSP or CPU power and causes more latency. K-System RMS meters are available from many manufacturers while TC Electronic's even more revolutionary LM5 true loudness and loudness history meter is now available for Pro Tools|HD.

According to Katz, mixing engineers would be better off dispensing with meters altogether and using their ears instead! As he explains, "having calibrated monitor gain is just as important as metering peak-to-average ratios. It is possible to mix an entire album 'blind', without any metering at all, yet never overloading the digital system! All you need to do is set a sufficiently high monitor gain (e.g., 83 dB at −20 dBFS RMS). When mixing this way, mix engineers can mix using their ears without the arbitrary constraints or influence of meters. The mixes which result will likely have a better crest factor (peak-to-average ratio) than typical mixes made while watching meters and, later on, in mastering, should produce louder masters with far less sonic compromise."

As I pointed out, "With the next (and even some current) generations of digital audio broadcast and consumer replay equipment, listeners will be given options to select dynamic ranges for replay. Consequently, music that has been squashed dynamically to sound louder will not sound louder any more, but will sound harsher in comparison with music that has not been so badly 'mangled.'"

An AES paper written by Thomas Lund that addresses these issues can be downloaded at www.tcelectronic.com/media/lund_2006_stop_counting_samples_aes121 .pdf.

More of Bob Katz's views on metering are explained in documentation for the TC Electronic Finalizer, which can be downloaded at www.tcelectronic.com/ media/katz_1999_secret_mastering.pdf.

Other articles of interest can be downloaded at www.tcelectronic.com/ TechLibrary.asp.

TL MasterMeter Oversampling Metering

It has become fashionable in the music industry for clients to demand the loudest possible mixes—the theory being that, these will stand out among the crowd. The problem is that, when levels are at or near 0 dB FS for much or all of the time, it is almost certain that when these recordings are played back on consumer CD players, significant amounts of fatiguing distortion will occur when the digital audio is reconstructed into analogue audio. This can happen even when the highest samples are kept just below 0 dBFS, because the reconstructed analogue waveforms can raise higher in level in between these sample peak levels—and will then be clipped by the analogue playback equipment.

The important point to note here is that, the sample values in digital audio do not represent the peak level values of an analogue waveform reconstructed from these samples—peak levels will frequently exceed these values. Also, if you make a test CD and play it back on a high-quality CD player in your recording studio, it may have high-quality analogue electronics with sufficient headroom to successfully playback levels that exceed 0 dBFS without producing audible distortion or fatiguing effects—as a result, you may be seduced into believing that there is no problem with your mixes!

The standard meters in Pro Tools will not alert you to this situation—they just measure sample peaks. What you need is a meter that simulates what happens during the reconstruction of the analogue waveform and alerts you if clipping would happen after reconstruction. If you know where and by how much the clipping takes place, you can go to these places in your mix and adjust levels to avoid this clipping.

There is a plug-in that will do all this: the TL MasterMeter TDM and RTAS plug-in is now included with all Pro Tools systems. This allows engineers to compare regular and inter-sample peaks over time and make appropriate adjustments to counteract any clipping due to inter-sample peaks. Over-sampled peak meter like this, which simulates the reconstruction filters used in digital to analogue converters, is an essential tool to use—especially with mixes that have been processed to be particularly loud or "hot."

Using the TL MasterMeter

The TL MasterMeter has two separate meters, one that shows the standard signal level and another that shows the oversampled signal level. Because the oversampling process can create levels above 0 dB, this meter shows an expanded scale from 0 dB to +6 dB. It is easy enough to use: simply insert TL MasterMeter on a Master Fader track and play the entire session back from the beginning to the end to check your final mix.

You can easily keep an eye on these meters while you play the mix and if you see any clipped signals, just stop playing back immediately and look for the ways to fix the problem. This is fine if there are just one or two "overs" during the mix. But if there are a lot of them, you will find it quicker to let the whole mix play through and then look at the historical list of events that is created in the two "browsers" provided for signal clips and oversampled clips. Because the timecode for each event is listed, you can quickly go to each location where there is a problem and fix it.

The Signal Clip Events browser displays historical clip events and has columns that show the relevant time code for the beginning and ending of each clip event. With stereo tracks, the first column shows "L" or "R" to indicate if the left or right channel has clipped. The Min and Max values in this browser will always be zero, unless the Clip level is set less than zero. At the bottom of the browser, the Peak field displays the highest dB value of the audio signal received so far.

The Oversampled Clip Events browser displays oversampled clip events, "historically" (sequentially). The amount of potential clipping in excess of 0 dB is also displayed. The columns show the timecode for the beginning and ending of each clip event, as well as the minimum and maximum clip values created after oversampling. With stereo tracks, the first column shows "L" or "R" to indicate if the left or right channel has clipped. At the bottom of the browser, the Peak field displays the highest dB value of the oversampled audio received so far. The Events field below the browser shows the total number of clip events in the oversampled audio signal.

In the accompanying screenshot, you can see what was displayed when I inserted the TL MasterMeter plug-in on the Mono Master Fader used in the previous example. An audio track was playing back a short recording of a 0 dBFS, 1 kHz test tone through this mono Master Fader. So, the Signal Level meter shows the level to be 0 dB. However, the clip light on the Oversampling Level meter lit up almost immediately, and there were 8 consecutive oversampled clip events displayed in the Oversampled Clip Events browser, all hitting Full Scale.

Phase Measurements

When you are working on stereo mixes, it is always useful to keep an eye on the phase correlation between the left and right channels to make sure that you have good mono compatibility. Of course you should be checking your mixes by listening in mono from time-to-time as well, but convenient availability of phase scope and a phase correlation meter is definitely a bonus! The DigiRack PhaseScope is ideal for this purpose, as it provides signal level and phase information for stereo tracks in both TDM and RTAS formats.

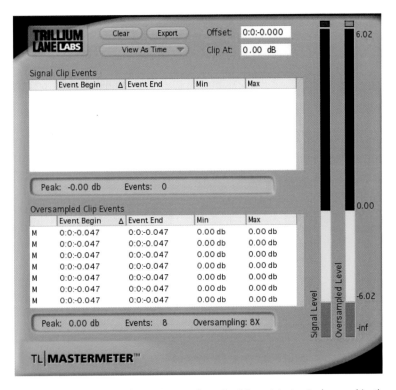

Figure 7.9 TL MasterMeter plug-in inserted on the Mono Master Fader used in the previous example. This shows an "over" with 8 consecutive oversampled clip events hitting full scale—providing a possible clue as to why the clip light on the Audio track that was playing the 0 dB FS test tone was lit.

Lissajous Figures

On a conventional oscilloscope, the left signal is sent to the vertical Y-trace and the right signal is sent to the horizontal X-trace inputs. This produces curves known as Lissajous figures on the graph. A curve with a slope that raises from left to right at a 45° angle shows that the left and right input signals are in phase. A curve with a slope that raises from the right to left at a 45° angle shows that the left and right input signals are out of phase.

DigiRack Phase Scope

The DigiRack Phase Scope is rotated 45° to the left, so that the Lissajous figures raise vertically to indicate in-phase components and horizontally to indicate out-of-phase components. In other words, when the curve traces vertically, it shows that the combined left and right input signal is in-phase and when the curve traces horizontally, it indicates that the left and right input signals are out-of-phase.

Typically, stereo recordings produce a random pattern that is taller than it is wide. A vertically oriented pattern, produced by strong correlation between the right and left channels (approaching mono, which would produce a straight vertical line) indicates good mono compatibility. If the pattern runs horizontally, this indicates a lack of correlation between the two channels, which would lead to cancellations between out of phase components if the left and right signals were combined to mono, and should obviously be avoided if good mono compatibility is required.

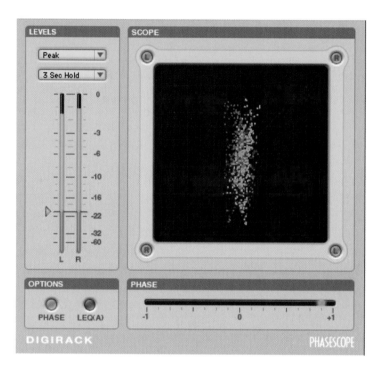

Figure 7.10
Phase Scope.

Phase Meter

Below the phase scope, there is a Phase Meter display. A phase meter, also known as a correlation meter, indicates the degree of similarity, or correlation, between the left and right channels. When the audio in the left and the right channels is similar, the meter moves towards the right. If the audio in the left and the right channels is identical, the correlation would be +1 and the meter would be positioned all the way to the right. When the left and the right channels are dissimilar, with lots of different audio material in each channel, the meter moves towards the left. If the audio in the left and the right channels is exactly out of phase, the correlation would be −1, and the meter would be positioned all the way to the left. At the center or zero position, the signal is a perfect stereo image. Most recordings have phase correlations somewhere between 0 and +1. Short excursions into the negative area at the left are not necessarily a problem, but anything more than this could indicate mono compatibility problems.

> ### Note
>
> If you widen the stereo or reverb width for your audio, the phase correlation will move towards the left side as the left and the right channels become "wider," i.e., less similar.

Leq(A) Meter

Leq(A) is the equivalent continuous A-weighted sound pressure level determined over a measured time interval that takes account of fluctuations in level within this time period to produce an average value. This average unfluctuating Leq(A) value is used when specifying loudness levels that must not be exceeded in order to comply with the regulations, for example, or when recommending monitoring levels. The A-weighting filter approximates the loudness sensitivity of the human ear.

You can switch the Phase Meter display to an Leq(A) Meter display that shows the *true weighted average* of the power level sent to either of the two stereo channels. By default, both channels are selected for display and colored green indicates that they are selected. You can deselect the left or the right channel by clicking on the "L" or "R" buttons at the corners of the display.

The Leq(A) Meter display shows a floating average for the level over the interval chosen in the Window menu. For example, with a setting of 2 seconds, the display shows the average value for the most recent 2 seconds of audio playback. A popup "Window" menu lets you choose the length of time (between 1 second and 2 minutes) that the signal is measured over before an average value is calculated. You can also choose infinite mode, which constantly averages the signal instead.

Enabling "Auto Reset" causes the start time of the Leq(A) measurement window to be automatically reset whenever playback starts in Pro Tools. Enabling "Hold on Stop" causes the Leq(A) measurement window timer to pause when playback stops and resume when playback starts again.

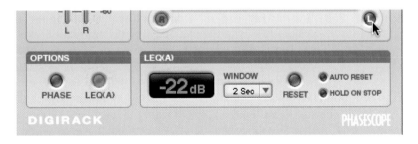

Figure 7.11
Deselecting the Left channel for Leq(A) metering.

Level Meters

To the left of the Scope display, there are meters for the left and right input signals. There are two popup selectors above the meters, one for peak hold options and another for the meter type.

Meter types include Peak, RMS, and Peak + RMS. Peak (default meter type) uses the same metering scale that is used in DigiRack EQ III and Dynamics III plug-ins. RMS (Root Mean Square) uses a proprietary "true" RMS metering scale. This "true" RMS meter scale is not same as the AES 17 RMS scale. For a sine wave with a peak value of −20 dBFS, the "true" RMS meter will show a value of −23 dBFS. For the same sine wave, AES 17 RMS meter will show a value of −20 dBFS. Peak + RMS uses a multi-color display to differentiate the two types of metering with Peak metering shown in green and RMS metering shown in blue.

VU (Volume Units) metering is provided as an option and this uses AES standards to indicate the signal levels. Various broadcast metering standards are also provided, including BBC IEC-IIa, Nordic IEC Type I and DIN IEC Type I, which all use reference calibrations of −18 dBFS. There is also a meter scale specially developed for Avid's VENUE systems that uses a reference calibration of −20 dBFS.

Audio Track Faders

When you play back an audio track in Pro Tools, you are reading a file from a disk drive. If you set the fader at the unity gain position, neither boosting nor cutting the signal level, the meter (or meters on stereo or multi-channel tracks) will show a moderate signal level—not too low to have problems with the signal-to-noise ratio and not too high to cause clipping. Assuming that the audio was recorded correctly, this will be the case.

If the audio was recorded at too low a level, then you will probably hear background noise if you boost the fader a lot, and there will be little that you can do about this—which is why it is so important to get recording levels right in the first place. You may find it convenient to apply a gain change using AudioSuite to bring up the level of any tracks that are too low to avoid having to use ridiculously large boosts on these tracks with the faders. Pro Tools|HD faders will allow you to boost levels by up to 12 dB.

If the audio clipped while it was being recorded, you will probably see the track's red overload indicator light up. Obviously, if the audio is too badly clipped it will be unusable. But if there is just an occasional overload, especially if it is inaudible, the best way to deal with this is to insert a Trim plug-in and drop the level coming from the audio file on disk by half a dB or a dB or whatever, until the clip indicator no longer lights up. Then you can concentrate on making sure that no additional clipping is taking place because of the way you are operating your mix session—without your attention getting distracted by the clip indicator. This procedure does not fix any clipped audio, it just helps to ensure that you don't make the problem worse by adding further clipping to that audio during your mixing session. If you do add further clipping, perhaps when using a plug-in, you will know that the red clipping indicator is now lighting up for a good reason—to warn you that you have clipped the signal again!

Master Faders

Master Faders control the output levels of the output and bus paths. Post-fader inserts are provided, so that you can also apply effects processing along with the level control—for your main mix; headphone or cue mixes; and for any stems, effects sends, or other signal-routing applications. Master Faders can also be used to meter a bus or hardware output to guard against clipping.

> ### Note
> Unlike inserts on audio tracks and Auxiliary Inputs, Master Fader inserts are post-fader. While plug-ins are inserted, you will affect the processing if you alter the Master Fader level; because changing the Master Fader level will change the level being sent to the plug-in. Also, don't forget to ensure that there is no clipping occurring in any plug-ins that you have inserted onto the Master Fader.

Although you do not have to use Master Faders in Pro Tools, you should normally choose to use these to get the best results. Master Faders do not use any DSP, so you won't suffer any performance "hit" by using these; and, most importantly, Master Faders allow you to trim your final output level to avoid clipping the DAC (Digital-to-Analogue Converter) or 24-bit digital output when your mixed signals leave the Pro Tools mix environment. Also, by observing the Master Fader's meters, you can see if any clipping is taking place due to summing the individual channels.

If clipping is taking place, which will cause the Master Fader's red overload indicators to light up, you could lower all the individual channel faders to fix the problem, but it is much more convenient to simply lower the Master Fader and play your mix again until you are sure that you have corrected the problem. The important point here is that you will not lose any quality by lowering the Master Fader by a few dBs to adjust the gain at the output stage, because there is more than sufficient dynamic range available within Pro Tools|HD with its 48-bit processing capabilities.

Monitor Controllers

Of course, if you keep lowering the Master Fader, you will reach a point where you will affect your mix adversely. Some of you may be tempted to connect the outputs from your Pro Tools hardware interface directly to a pair of powered monitors or to a stereo amplifier driving a pair of passive monitors. In this case, the only convenient way to lower the listening level would be to lower the Master Fader level. But this would be crazy because you would never be able to get your mix levels correct if you always had to lower the Master Fader to make the playback volume low enough to avoid discomfort, or deafness! The solution here is to use a monitor controller that provides a volume control. Popular models such as the Mackie Big Knob, and similar models from PreSonus and SPL are very affordable and usually include talkback and headphone monitoring features as well.

Mixing Precision and Master Faders

The Pro Tools mixer is designed to allow the faders to be lowered in level without any loss of resolution. Normally, in a 24-bit system, as you lower the fader, you are using less than the available 24 bits resolution. However, the Pro Tools mixer uses DSPs on the audio cards that temporarily use 48-bit calculations when mixing signals together. So, even if you lower a fader almost to the bottom of its travel, you still have 24 bits of resolution available.

Using 48 bits for the DSP, with 56-bit registers to hold interim calculations, provides more headroom (and "foot-room") than you are ever likely to need. This means that when the input signals are summed together onto the mix bus in Pro Tools, these signals can never clip—even with all the channel faders set to the + 12 dB maximum gain—because there are more than enough bits and headroom available internally to avoid this.

Similar to analogue mixers, the Pro Tools mixer is comprised of individual input channels and a summing stage. At the input stage, each channel's 24-bit word is multiplied by 24-bit gain and pan coefficients to create a 48-bit result. The new 48-bit word contains the original 24 bits "shifted" lower in the 56-bit register to allow for the headroom and "foot-room" below unity gain, enabling channels to be turned down without losing precision. Specifically, it's possible to pull any channel fader down to −90 dB and its signal still retains 24 bits of precision. As channel faders are pulled down, there is a loss to the lower bits of the newly extended 48-bit word, which represents signals down to about −240 dB—but a full 24 bits of precision is maintained down to −90 dB.

The situation at the output side of the summing mixer, where audio is sent via a digital output or onto the TDM bus (or when you sub-mix signals to an Aux Input) is different—clipping can occur here. But that's what Master Faders are for—they allow you to trim your final output level to avoid clipping the DAC or 24-bit digital output when your mixed signals leave the Pro Tools mix environment. This is analogous to an analogue console mix bus, where you trim the master bus with a Master Fader to avoid clipping the output circuitry in the console.

> ### Tip
> This is why you should always be using a Master Fader to scale the output level of any mix summing point onto a bus or output.

For a detailed discussion of these topics, download the "Pro Tools 48-Bit Mixer" Digidesign Technical White Paper by Gannon Kashiwa from the Digidesign Technical White Papers section of the Avid Knowledge Base website. This provides detailed information about how the Pro Tools mixer operates and demonstrates its summing characteristics, explaining how a 48-bit "clean" mixer functions within the 24-bit TDM (Time Division Multiplexing) environment. It provides "behind-the-scenes" information about mixing and summing in Pro Tools and sheds light on various myths about mixing "in the box" with Pro Tools, that will give you a much better understanding of the mechanics of summing signals.

Using a Master Fader to Prevent Clipping on a Bus

A Master Fader can be used to control the level of the audio on an output or outputs. It can also be used to control the level on a bus. This is actually a very important application of Master Faders, which deserves to be more widely used.

For example, it is very common to sub-mix drums onto an Auxiliary Input channel so that you can control the overall level of the kit using one fader and so that you can insert effects such as compression or reverb across the whole kit. The problem here is that, sometimes the combined signals from all the audio tracks that are feeding the Auxiliary Input can cause the level on the bus to clip, even though the levels of individual audio tracks do not clip. And you can't use the fader on the Auxiliary Input (or on a Send from this) to prevent this clipping.

As you can see in the accompanying screenshot, the clip lights are red on the Aux Input and on the Send to the reverb. Lowering the Auxiliary Input fader or lowering the Send fader will not prevent this clipping, as these faders simply control the amount of the clipped audio that is sent to the mix bus and to the reverb device. If you try doing this, the meters will still show that the audio on the bus is clipped.

You could lower some or all of the individual faders until the summed level at the Auxiliary Input no longer clips, but you run some risk of disturbing the balance you have carefully set up between the individual tracks, and this is not as convenient as lowering one fader.

Fortunately, there is a solution to this problem: using a Master Fader to monitor and control the bus levels. Simply insert a Master Fader and assign this to the bus that you are using to route the audio tracks into the Auxiliary Input to form your subgroup.

Figure 7.12
A typical sub-mix of drum tracks onto an Auxiliary Input with a Send used to add reverb to the whole kit. Note that the clip lights are red on the Auxiliary Input and on the reverb Send.

Figure 7.13 Using a Master Fader to monitor levels on the bus used to sub-mix the drums onto an Auxiliary Input. Note that the levels on the bus are clipped, as revealed on the Bus Monitor meters and on the Auxiliary Input meters and the Send meters. Lowering the Send level has not prevented clipping, and lowering the Aux fader level would not either.

Whenever you see the clip light turning red on a Master Fader that you are using as a Bus Monitor, you can lower the level of the Master Fader to reduce the level on the bus until it stops clipping. You could also insert a compressor or limiter on this Master Fader to "tame" the peaks if you like.

Figure 7.14
You can lower the Master Fader used to monitor levels on the drums sub-mix bus to prevent clipping, as has been done here.

Inside or Outside the Box?

A big decision for some is whether to mix completely "inside the box," keeping everything digital inside the Pro Tools, or whether to mix some or all of the tracks "outside the box." This could involve using a small external mixer or a specialized "summing mixer" with 8 or 16 tracks, to take the most important tracks with the rest mixed inside Pro Tools, or it could mean routing all the tracks from Pro Tools individually into a large-format external console such as an SSL or Neve.

Top UK recording engineer and producer John Leckie says that he finds summing mixers useful with Pro Tools rigs to provide hands-on faders to allow the guys in the bands he works with to take part in the mixing sessions. He also values the additional facilities that some of these units provide.

For another opinion about this, I asked George Massenburg who is one of the leading recording engineers and producers in the US. He told me "The original DAW mix always sounds best to me. Mixing in the box has more detail and is less colored."

> **Note**
>
> For more information and a discussion about summing mixers, read Appendix 1 on this book's companion website. This contains an article that I wrote for Pro Sound News Europe all about summing mixers called "Summing to talk about!"

The "Tools of the Trade"

Mixing primarily involves adjusting the levels of the various instruments, voices, and other mix "elements" such as sound effects that make up your mix. Beyond simple level control, there are many other processes that may be involved in creating a finished mix. Pro Tools provides you with, or lets you access, all the "tools of the trade" that you will need to make, even the most complex mixes. These "tools" include the wide selection of signal processing plug-ins supplied with your Pro Tools system, third-party plug-ins that you can buy, and any external hardware effects that you hook up to your system.

Level Balancing

If you come across older books about sound studios or if you work for the BBC, you will hear the term "balancing" used to describe what most people call "mixing." "Balancing" is actually a very good way of describing the most important thing that you will do when you are mixing a multi-track recording. This is to decide how loud or soft the various elements of the mix should be—which you do by "balancing" the level of each instrument or sound against the others, moving each one up or down in level until the mix sounds the way you want it to.

> **Tip**
>
> The balance in levels between the various mix elements is probably the single most important aspect of most mixes. Get this right and you are a long way along the path towards a great mix!

One decision to make is whether to keep the level balance similar to those that you would expect to hear in a "live" situation or whether to balance the instruments in such a way as to create unique perspectives that would not normally be heard in "live" playing situations.

If you are recording a string quartet playing conventional repertoire, for example, you would record this in a suitable room (in a recording studio or at a "live" venue) in stereo to capture the overall effect, possibly with individual microphones on each instrument so that the balance between these could be fine-tuned at the mixing session. A jazz group could be recorded in a similar way, with the aim of delivering the most realistic, "believable" sounds to the listener. The recording engineer would aim to capture the sound of the instruments arranged as they typically would be onstage and as heard by a member of the audience located centrally in the listening area.

With many genres of popular music present, on the other hand, it is more likely that you will create balance between the instruments that you would not normally hear when these instruments play together in a room. For example, you might have the drums very loud in the mix with a brass section playing riffs much more quietly in the background. Or you might have a quiet acoustic instrument such as ukelele turned up loud enough in the mix to be heard as a solo above a full electric band with keyboards, guitars, bass, drums and, so on all playing at the same time. Or maybe you could have some percussion instruments such as triangle or tambourine playing at a very low level in the mix—but just sufficiently loud to be heard and have a useful effect.

Panorama (Pan) Positions

Positioning the mix elements within the stereo "panorama" is another important part of the "art" of mixing. Panning is the subject of much debate and ultimately comes down to your own creative choices. To create a satisfying mix, your goal should be to arrange the mix elements within the stereo sound field such that they work well together without obscuring each other. It is also a good idea to aim for some kind of symmetry so that you don't end up with mix elements going on to the right or to the left. For example, you might pan one percussion instrument to the left and another to the right. Or you might have an acoustic piano on the left and an organ or electric piano on the right. Of course, there will be times when the way the musical arrangement develops means that this is not always possible, but it is generally a good goal to aim for.

Tip

If you don't have another instrument in the arrangement that you can use to create balance, you might pan a delayed version of one of the mix elements to the other side. This can be very effective with guitar or percussion parts, for instance.

Some early stereo mixes used radical pan positions—such as bass and drums hard left; keyboards and guitars, brass, or percussion hard right; and vocals in the centre. This was sometimes due to technical limitations in the equipment used. On other occasions this was done as a result of the creative choices made by the engineers and producers. Mixing in this way has the advantage of leaving lots of room for the vocal to be the absolute centre of attention, with the other instruments very much in a supporting role—which is ideal if you have a great singer and a great song.

I have recently been listening to some early stereo mixes of late 60s pop/soul songs—hit records such as "Goin' Out of My Head" by Little Anthony & The Imperials and "Hey Girl" by Freddie Scott—which typically might have the rhythm section panned hard left; backing vocals, brass, and strings panned hard right; and lead vocals in the centre. This way of mixing had the advantage that the lead vocal was easy to hear—loud and clear and with no competition from other instruments occupying this central space. This technique is still occasionally used today. One example of this is a song called "Empty" on Ray Lamontagne's 2006 album "Till The Sun Turns Black" which has drums panned left and acoustic guitar panned right with the vocals in the centre.

Nevertheless, it is much more usual today to have the lead vocal, the bass guitar, and the bass drum all dead centre; with the snare sometimes centre or maybe, as with the hi-hat, panned left or right; and with the kit overheads capturing a natural stereo spread of all the drums, including the cymbals. Talking about panning the drums, you may wish to consider to pan these the way the audience would hear them or as the drummer would hear them. There are arguments in favor of both ways. If you pan from the audience perspective (this is often what the audience expects), then the music probably sounds the most natural. On the other hand, panning from the drummer's perspective lets you listen from the drummer's seat on-stage.

Guitars might typically be panned half left or right, with keyboards spread a little further out—but don't be afraid to experiment here. It can sometimes be very effective to use an auto-panner (or Pro Tools automation) to move instruments around as the music plays—like a wandering percussionist or acoustic guitarist might do on-stage.

It is also worth keeping in mind where the mixes are going to be played. If you are aiming the music to be played at the audiophile with a great stereo system, allowing the audience listening in the sweet spot, then you can make good use of extremes of panning. On the other hand, if the mixes are intended for a dance club, even if this has some kind of stereo speaker system installed, you can't count on dancers being in any kind of sweet spot for too long, so you should keep the important rhythmic elements of your mix much closer to the center.

Although the stereo panorama that you can create using just two speakers basically allows you to move the mix elements along a horizontal plane running from far left to far right; it is also possible to create a sense of depth, and even of height, using careful choices of delays and reverberation. A word of warning here—you do need to be careful about your choices of pan positioning if you are aiming for good mono compatibility. If you have two guitar parts playing in similar frequency ranges, for example, you can make these clearly audible as individual parts in a stereo mix by panning them hard left and hard right. Unfortunately, when you listen in mono, it will often be much more difficult to distinguish one part from the other as they will both be heard playing in the same place.

Forewarned about this (by having checked in mono) you can try other ways of distinguishing the two parts, perhaps by EQing the two parts differently and then narrowing the distance between the two parts in the stereo mix (so that there will be less of a difference when you "collapse" from stereo to mono) before checking again in mono to see how well this works.

> ### Note
> If you want to create a mix with good mono compatibility, you should regularly check how the mix sounds in mono and adjust the mix accordingly.

> ### Tip
> If you pan mix elements while monitoring in mono, you may be able to find an optimum pan position in which each element sounds more audible when summed to mono.

Of course, if you are working in surround, you will have many more choices when it comes to panning: 360 degrees around the listening position and all points within. This is why surround mixes can sound much more natural, because it is much easier to achieve separation between the different instruments in the mix.

Panning Synthesized and Sampled Sounds

By default, most synthesizer and sampler sounds are set up to work in stereo, panned hard left and hard right. One thing to watch out for is the trap of including these synthesizer parts in stereo in your mix—typically with the dry sound panned hard left and a chorused version panned hard right, although sometimes the second channel contains synthesized sound components as well. You should almost always reduce the width of the panning, and place these sounds more appropriately within your mix. Better still, just use the original synthesized or sampled sound in mono (or stereo) without its reverb or other effects so that you can position this exactly as you wish in your mix and apply the higher quality reverb and other effects that are available in Pro Tools. Your goal should be to find pan positions for the synthesized instruments that work in the context of your mix—and the sound designers who set these sounds up for your synthesizer or sampler could not have panned these suitably except by lucky accident. So don't go with the defaults—pan the synthesizers carefully yourself!

Pan Laws

As Avid explains: "There are a variety of ways that analog and digital mixers handle *stereo pan law*, the level calibration when a signal is set to center. For example, a track set at unity gain that is panned center will output a level of −3 dB to both the left and the right speakers. But when you pan it hard left, it will output at unity gain."

Variable stereo pan depths are available in Pro Tools 9 from the Session Setup window. The default setting for this is −3 dB in Pro Tools 9 (it was −2.5 dB in the earlier versions) and it can be adjusted from −2.5 to −6 dB.

The "pan law" (as it is normally termed) compensates for the 3 dB boost that happens when panning sounds from the extremes to the centre position by subtracting 3 dB from the level in the centre. This keeps the level of the panned elements the same when they are moved from the extremes to the centre position.

When you move a sound into the centre position, it is more likely to be masked by many other sounds that are likely to be positioned in the centre. So, you may prefer to use the −2.5 dB setting which will boost the level when panned centre by 0.5 dB.

The pan law value used in many mixing consoles, particularly for surround sound, is −3 dB, so the new default setting generally makes much more sense by falling in line with this industry standard.

The pan law value used in many older UK-designed analogue mixing consoles is −4.5 dB for stereo or −6 dB for full mono compatibility, so it is good to see that these options are now available for Pro Tools.

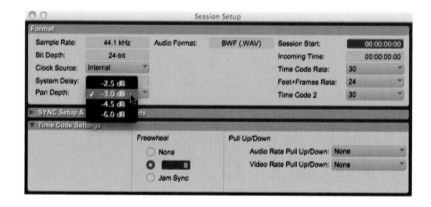

Figure 7.15
Session Setup window showing Pan Depth options.

Using Compressors to Smooth Out Level Changes

Once you have panned your mix elements and achieved a rough balance, you should listen to the whole piece of music from the beginning to the end. If you notice that any of your tracks are uneven in level throughout the piece, you can always use the volume automation in Pro Tools to smooth out these level changes.

Of course, this can become a very time consuming task if the singer, guitarist, bass player, or whoever is not experienced enough to sing or play very evenly so that their levels are constantly changing throughout the piece. You can often get quicker results by using compression to smooth out the level changes, then setting the overall level for the track using a combination of the makeup gain in the compressor and the fader level in the mix.

Just to remind you here, a compressor reduces the highest peaks in the audio, making the sound more even. When the peaks have been reduced, you can raise the average level of the audio without the peaks exceeding the maximum level that the system can handle without distortion.

> **Tip**
>
> You may not need to use compression on the bass, guitar, drums, or other instruments if you are recording experienced studio musicians. I have often been amazed at how evenly the top studio musicians here in London are able to play—with wonderful control over the dynamic levels that they are producing from their instruments.

> **Note**
>
> Compression should always be applied carefully: it is all too easy to take away the natural "openness" of a recording when you apply too much compression—"squeezing" the "life" out of the sound. And when you have chosen the compression settings that you think are working for you, make sure to compare the sound without the compression again to comfirm that you are really happy with the results.

Using the Gain Plug-in to Smooth Out Level Changes

Another way to smooth out unwanted level changes is to select any audio that is too loud and apply a gain change using the AudioSuite Gain plug-in. This way, you avoid all the unwanted side effects of compression such as the "breathing" effect that you sometimes hear when the compressor kicks in and then releases. It is very easy to do this for just one or two notes or sounds that are too loud and this can be a useful technique to use in combination with other techniques for smoothing out the levels.

Figure 7.16
AudioSuite Gain
plug-in.

De-essing

Another problem that often becomes apparent during a mix session is sibilance on the vocals that becomes even more noticeable when you raise the level and maybe boost the mid and upper frequencies using EQ to make the words easier to hear. In this case, it is just the so-called "sibilant frequencies"—which are higher frequencies that are prominent in the syllables "s" and "f"—that need to be compressed because they are too loud.

The trick here is to use a "de-esser." This is a compressor designed specifically to alleviate problems with S's and other sibilant sounds by applying fast-acting compression just to the sibilant frequencies. Such a device uses EQ filters in a "side-chain" to isolate the sibilant frequencies so that the compressor will only affect the levels of the problematic frequencies, leaving the rest unaltered.

> **Tip**
>
> When applying level changes to balance overall track volume levels, it is also possible to apply individual gain changes just to the sung notes that are suffering from sibilance problems. And, again, you can apply these changes in real-time using Pro Tools volume automation or as an "off-line" process using the AudioSuite Gain plug-in. If there are just a few sibilant notes that need adjusting, or if you have a lot of patience, this can be a more effective technique to use than using a de-esser—which is a more automatic process that does not pay such detailed individual attention to each sibilant note.

Creating Dynamics within the Mix

Often you will realize that some of the mix elements ought to get softer or louder during the mix to create and maintain interest. For instance, if you have a solo instrumental in the mix, such as a guitar player who improvises throughout the piece, you could turn this up louder wherever nothing much else is happening in the music and make it quieter whenever there are lots of other mix elements vying for the listener's attention.

Also, some of the constant mix elements could be muted in particular sections so that they make the sections, where they can be heard, stand out better. So, for example, you might record a guitar or keyboard playing a rhythmic figure throughout the piece, and then try muting this in the verses—just using it in the choruses.

> **Tip**
> Don't hesitate to record or add a new material during the mix session if you feel that something is missing musically from the arrangement. When the singer is not singing, for example, you might want a guitar or sax to play a solo that was not thought of during the track-laying sessions; but you now realize that this would be the "icing on the cake" to complete the recording.

Frequency Balancing

You can balance the frequencies in a recording either by the choices that are made in the musical arrangement or using the tools available in the recording equipment that you are using.

Musically

Having different instruments play in different frequency ranges (i.e., in different octaves) usually helps to make a more satisfying sound than if several are playing in the same range, where they may clash with each other. The choice of which instrument plays where and when may be made by the musicians or bandleader, or it may fall to the songwriter, composer, arranger, or producer.

You should always aim to create a good frequency balance in the musical arrangement, which is why it can really help to employ the services of a good musical arranger to work on your music production, if you do not already have someone with these skills involved.

Technically

You can also use EQ to adjust the frequency balance of any of your mix elements so that you create a more balanced range of frequencies within your mix.

For instance, electric or acoustic guitars can have quite a lot of low-frequency content that can interfere with the bass guitar or other low-frequency instruments. Filtering out the lower-frequencies or reducing the level of the low-frequency content can make the guitar parts blend into the mix much better.

For example, I wanted to blend in a recording of a Martin HD28 acoustic guitar with a mix that also contained bass guitar, drums, electric guitar, tenor banjo, mandolin banjo, vibes, percussion, and vocals. I used the following EQ settings on the Martin guitar to achieve the result I was seeking:

Subtracting (–) 3.5 dB at 50 Hz took some of the "muddiness" away from the bottom end, allowing the bass guitar and bass drum to be heard with less interference from the acoustic guitar's low frequencies.
Adding (+) 1.5 dB at 100 Hz gave the guitar more "body" or "fullness", helping the sustaining chords to act as a "pad" sound.
Adding (+) 1.5 dB at 3.2 kHz "brightened up" the chords in the mid range, making the tonal detail and the pick strokes much easier to hear.
Adding (+) 1.5 dB at 6.3 kHz let the guitar "cut through" the mix with much more clarity.

Remember that this is just an example. Other than by coincidence, these settings are not going to work for you, because the guitar will be different, the player and pick will be different, the microphone and preamplifier will probably be different, the room will be different, the other instruments in the mix will be different, and so forth. So you will always have to find your own settings by trial and error, although you could always try the above settings as a starting point and you just might get lucky!

One way to look at EQ is to consider the boost/cut controls for the various frequency bands simply as volume controls for those bands—which is, essentially, what they are. When you think of EQ this way, it is easy to see that lowering the volume of particular frequency bands allows those that are not cut to stand out more prominently. So, instead of boosting the mid-range, you could simply cut the highs and lows and the mid-range will be emphasized. Conversely, if you want to emphasize the highs and lows, you can cut the mid-range. This approach also helps to avoid causing clipping, which can happen when you raise levels.

Reverb

Reverberation occurs naturally in rooms, halls, and most other acoustic spaces and is composed of myriads of reflected sounds bouncing around the acoustic space, reflected by walls, floors, ceilings, and any other hard, reflective surfaces, such as glass windows or screens within the space. Sound is absorbed by soft furnishings, curtains, carpets, human bodies, and so on. So, in real rooms, the reverberation will change according to how many people are present, and even whether the curtains are drawn or not. Harder surfaces reflect sound and if there are many small surfaces that reflect the sound in different directions, these will make the reverberant sounds more diffuse. To make a small room sound larger than it is, it is a good idea to use special diffusers.

In classical, orchestral, and some jazz and other recordings, the goal is to capture the natural reverberation of the room along with the sound of the instruments. The so-called "ambience" microphones can be placed wherever is appropriate in the room to capture this sound and blend it in with the mix. Artificial reverberation is often added to recordings to enhance the natural reverberation that already exists or to create reverberation around directly injected electric or electronic instruments such as guitars or electric pianos, or around synthesized or sampled sounds.

> **Tip**
>
> On a vocal, you might use a long predelay setting with the reverb so that the words retain lots of punch and audibility, while gaps between words are enhanced by the sound of the reverb kicking in.

My first choice of reverb for Pro Tools|HD systems has to be Altiverb from Audio Ease. This is a convolution reverb that has the largest set of high quality impulse responses, taken from many of the world's finest rooms, halls, chambers, studios, and so forth, giving you access to some of the best acoustic spaces from around the world. I always use this as my primary reverb. I have compared all the other reverb plug-ins for Pro Tools systems but Altiverb is way ahead of everything else. It was the first convolution reverb available for Pro Tools systems and has reached a stage of development that none of its competitors have been able to match.

Delay Effects

Delays can be used to produce obvious effects such as echoes, plus many other effects. For example, you can create a stereo image from a mono sound source using a short delay and set the width of this image by panning the delayed sound away from the source sound to create the required distance between them. If the delay is short enough, the two sounds will be heard as one sound with a wider image than the original source.

One of my favorite delay effects for Pro Tools is the Line6 Echo Farm plug-in, which simulates classic delay units such as the Maestro Echoplex—It is great for guitars, keyboards, trumpet, and other instruments.

For lead vocals, I recommend Sound Toys Echo Boy. This gives you a warm, high-end tape echo simulation, modeled after the Ampex ATR-102. It emulates 30 different echo styles including the Roland Space Echo and popular effects "pedals" such as the Memory Man and the DM-2.

One thing here to watch out for is that if the mix is summed to mono for playback on an old radio or TV, or whatever, the image widening effect, of course, disappears. So, if you have been relying on having particular instruments widely panned to prevent them from obscuring other elements within the mix and then when you sum to mono, your mix may suffer badly as a result. To counteract this tendency, make sure that you don't pan the mix elements too widely. Obviously, the closer you pan mix elements to the centre, the better the transfer to mono will be.

> **Tip**
>
> You should never be afraid to get rid of things that clash in the mix-recording or adding new material, if necessary. And, ultimately, you should consider a complete re-recording of the music with a new arrangement, if the mix just doesn't seem to work at all.

Noise Reduction

Sometimes, when you bring the level of a particular track up as loud as you want it to be in your mix, you will hear some background hiss (from a guitar amplifier, for example) that you definitely don't want to hear. This is where you might want to use noise reduction software (or hardware) to get rid of the hiss (or other noises).

I often record a Wurlitzer electric piano that has a very noisy preamplifier. This produces lots of hiss and 50-cycles mains hum that can often be very intrusive, especially on quiet passages, so I often find myself denoising Wurlitzer tracks.

Noise Gates and Expanders

An expander expands dynamic range rather than compressing it, so expanders can be used to reduce signal levels when the input signal falls below the selected threshold—making the difference between softest and loudest signals greater. This is called "downward expansion" and is just what you need to prevent any low-level noise that occurs in between "wanted" audio (with levels above the threshold) from being heard. This is why an expander can be used as a so-called "noise gate." When the input signal falls below the threshold, the expander makes the quiet sounds, so quiet that they cannot be heard—that is, the gate closes. When the audio input signal raises above the threshold again, the gate opens.

> **Note**
>
> Expanders can also be used to increase low signal levels when the input signal falls below a selected threshold. This is known as "upward expansion" and can be used to bring out quieter details in a track without increasing the signal levels of the other (louder) parts.

> ## Tip
>
> I find myself hardly ever using noise gates within Pro Tools because it is so easy to define a "silence" threshold and strip out any audio that falls below this threshold. Doing it this way avoids any possibility of false triggering, where an unwanted sound just above the threshold opens the gate. Of course, you still have to listen and check carefully to make sure that what you have defined as "silence" has stripped out everything you want to remove correctly.

Pitch and Time Correction and Manipulation

If the vocals or any instrument that is playing single notes (as opposed to double notes or chords) are out of tune, there are lots of software tools available to correct these. Sound Toys offers the fairly basic, but useable, Pitch Doctor plug-in, which offers both automatic and manual pitch correction. Auto-Tune is the best-known pitch correction software for Pro Tools. This has an automatic correction mode that is very easy to use and a manual graphical mode that is more advanced compared, for example, with Pitch Doctor.

Auto-Tune itself is now being rivaled by Melodyne, which has a more comprehensive feature set than the Auto-Tune and can handle multiple tracks of audio. Melodyne's new Direct Note Access technology will even allow you to access individual notes in chords and polyphonic audio—letting you see them, grab them, and edit them!

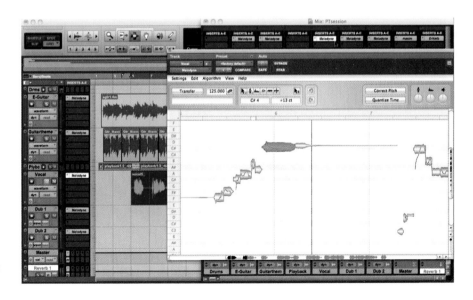

Figure 7.17
Melodyne plug-in in use in PT 9.

Harmony Processing

There are several plug-ins available from SoundToys, TC Electronic, and others that will let you create harmonies from a single voice. I can particularly recommend the Antares Harmony Engine. This lets you produce very realistic harmonies—you can even vary the "throat length"—and stunning artificial effects. This is a part of Antares AVOX 2 Vocal Toolkit, which includes ten vocal processors that will let you create just about any effect you could wish for from your vocal tracks.

TC Electronic offers its Helicon voice processor, which incorporates compression, EQ, voice modeling, and harmony processing plus effects like chorus, delay, and reverb. Its voice modeling techniques allow changes to be made to vibratos or to glottal attributes such as breathiness, rasp, or growl and changes to formants (warping or gender-bending) that can make a male voice sound female. Like the Antares plug-in, Helicon's harmony processing also allows multiple harmonies to be created, with pitch shifts to create choral effects.

The original Eventide Harmonizer is also available as a TDM plug-in. This lets you create a delayed and pitch-shifted copy of the input signal so that you can "broaden" the sound of instruments such as the Fender Rhodes using small amounts of pitch-shifting and short delays, or create more radical effects using greater pitch shifts and time delays that will be heard as separate harmonies to the original.

So, whether you need to correct or change the pitch of sounds or whether you are more interested in creating special effects, Pro Tools either provides or lets you access a wide range of tools that you can use.

Basic Pitch Shifting

Pro Tools does provide a pair of DigiRack plug-ins, Pitch Shift, and Time Shift that let will you select an audio file or region and change its pitch or its length, or change the tempo. For example, the guitarist might have played just one wrong note, say, a semitone lower than the correct note. Using a pitch shifter, you can select just this one note and move it to the correct pitch without having to trouble the guitarist to come back and correct this. Or maybe the mandolin player played an extra note in between two important melody notes and you have edited out the unwanted note, but the first note now doesn't sustain as long as it ought to. Just select the sustain portion of this note and time-stretch it to fill the gap from which the unwanted note was removed.

For small shifts of just a few semitones or beats per minute, the standard DigiRack non-real-time AudioSuite plug-ins work well enough. For larger shifts, you will get better results with third-party plug-ins such as Serato's Pitch "n" Time. Pitch "n" Time lets you alter pitch and tempo graphically or numerically and has a Varispeed Mode, where altering the tempo also alters the pitch.

Avid currently offers the highest quality pitch and time shifting algorithms in its X-Form plug-in, which lets you alter audio files by even-larger percentages with even fewer undesirable artifacts.

> **Note**
>
> Pro Tools HD also includes a real-time TDM plug-in called Pitch, which can be very useful for creating special effects.

Other Effects

There are times when you will want to use effects such as vocoders or other unusual effects processors such as the Aphex Aural Exciter and Big Bottom, all of which are available as Pro Tools plug-ins. In case you were wondering, the Aural Exciter processes audio to recreate and restore missing harmonics to revive the natural brightness, clarity, presence, and intelligibility of an original performance or to enhance the sound of any recorded audio. Big Bottom lets you create stronger, more powerful bass sounds with more bass density and sustain, using a dynamic EQ'ing process.

Or you might like to work with a multi-effects processor, which, as the name suggests, include lots of different effects. Early examples of these would include Yamaha's SPX90 and Eventide's H8000, which included reverbs, delays, tremolo, chorus, flange, and so on. The Eventide H3000 Factory multi-effects processor, also available as a plug-in, features the most comprehensive set of modulation effects that I have come across. One of the most useful multi-effects processors that I have seen also comes from Eventide: the Ultra-Channel incorporates three dynamics processors and a 5-band parametric EQ, along with Stereo Delays and Micro Pitch Shift.

Sound Toys Phase Mistress and Filter Freak plug-ins provide tempo synchronized modulation effects with lots of great presets for phase and filter effects, respectively. But my favorite multi-effects processor has to be Sound Toys Crystallizer. As it says on the website: "Inspired by the Crystal Echoes preset in the Eventide H3000, Crystallizer combines granular reverse echo slicing and retro pitch processing to create a huge range of radical sonic manipulations and classics with a twist. Use it to create synth-like textures from simple acoustic guitar rhythms, lush detuned echoes, or completely psychedelic pitch-shifted reverse echo effects. Great for drums, guitar, bass, sound design, electronic music, and just about anything else, Crystallizer is a truly unique and creative effects processor." I'd say that's about right!

Re-amping and Amplifier Simulators

If you have the equipment available and have the time and energy for this, you can always route your guitar (or any other) track out from your Pro Tools interface into a real guitar amplifier in your studio, put a microphone in front of its loudspeaker cabinet, and record this back into Pro Tools. This is called "re-amping" for obvious reasons. You can even go nuts and hire in lots of tasty vintage gear for the session—which costs a lot less than owning all the gear! But this can still be an expensive business both in the studio time needed and the costs of hiring in the gear.

Plug-ins such as Line 6 "Amp Farm" simulate amplifiers and loudspeaker cabinets very effectively and have lots of presets covering most popular combinations—and won't cost you a fortune! More advanced plug-ins such as IK Multimedia "Amplitube" or Native Instruments "Guitar Rig" also simulate microphones, boards full of foot-operated effects, and even racks full of studio effects. TC Electronic offers its Softube Vintage Amp Room, which faithfully recreates just three great guitar amps in a complete studio set-up with speaker cabinets and fully flexible microphone positioning. Vintage Amp Room has no extra gadgets or weird-sounding presets—it just accurately simulates classic Fender, Marshall, and Vox setups.

Avid offers its own re-amping tool called "Eleven." This is available as a TDM, RTAS, AudioSuite version for HD Accel systems, or as an RTAS and AudioSuite version for LE systems. It is also available in hardware as a rackmountable unit. Eleven has a great selection of Fender, Vox, Marshall, Soldano, and Mesa Boogie amp models plus convolution-based speaker combinations and mic modeling.

All these plug-ins are great for "re-amping" electric guitars, basses, or keyboards so that you can change the sound of these during your mix session. You can simply inject your electric instruments directly into the Pro Tools while recording, then build the sound you want for your mix at your leisure later.

> **Note**
> You can always change the sound of an instrument that was previously recorded through an amplifier or effects, but it works best if you have recorded a relatively clean sound with minimal effects—if you want to get the best results when re-amping and adding new effects.

It is always better if the guitar player plays through the amp that will be used for the track so that he or she can respond as a player to the sound of the amp. If it is a clean sound, the player will usually adjust his or her playing according to the sounds he or she is hearing from the amp. With a distorted sound, the player usually plays differently—maybe hitting more harmonics with the pick or holding the notes longer because they sustain longer with the distorted sound on the amp. And the way the player digs into the strings with the pick has a big effect on the way the amp responds. If the player uses thumb or fingers, the sound changes again with the amp responding differently to the way the strings are driving the input.

Working on Your Mix Session

It can be a good idea to let some time pass between your original track-laying sessions and the mix session. When you come back to it with "fresh ears," you may hear things in new perspectives, which can lead you to making a much more interesting mix than you had at first imagined.

But it can go the other way as well: you sometimes get the best mix by building everything up while you are recording, making edits, or adding effects as you go along and then tweaking and balancing everything to make the final mix immediately after the recording phase is finished and as you are ready.

"Housework"

One of the first things you will probably do at the start of a Pro Tools mixing session is to listen through to the tracks to see what's there—especially if you are mixing music that you have not heard before for a client, or if it is something you are coming back to from some time ago.

This is a good time to be doing some "housework." You could start by making sure that you have a separate backup copy before you do anything so that if you mess up in any ways, you can quickly get back to the original files. Then you should delete anything that you are certain that you will not need—secure in the knowledge that you can always restore the session from your backups if you make a mistake.

There might be lots of MIDI tracks that have since been replaced by real instruments, or vice versa. It is also quite likely that there might be several "takes" of the lead vocal or of a guitar or sax solo that may need to be "comped" to make one or more composite tracks containing the best phrases. There might be various layers of tracked-up backing vocals, rhythm guitars, or brass parts that could be mixed down to stereo to form more manageable subgroups or "stems."

There are probably lots of unused regions that you could easily get rid of to make it easier to see the ones you do want, although you will need to take care before deleting any whole regions to make sure that regions within these are not needed anywhere in the session. Again, if you make a mistake, you have your backups.

Track Ordering

You will probably want to establish the order of the tracks in the Pro Tools Mix and Edit windows the way you prefer this to be. I always have the Master tracks at the right of the mixer, with a click at the far left, drums next to this, followed by bass, keyboards, guitars, any other instruments, and then vocals. I know some people who would reverse this order. The main idea here is to have some kind of logical ordering of the tracks that makes it easier and quicker for you to work with these.

It can also be helpful to name some (or even all) of the regions, takes, or tracks more appropriately—especially if this will make things easier and more efficient for you when you get to the creative part of the mixing session. Don't forget that you can easily save alternative versions of any Pro Tools session if you want to keep the older versions, in case you need to revert to these.

Removing Unwanted Sounds

It is always a good idea to take a note of and then mute or delete any odd noises or sounds, clicks or pops, microphone stand rumble, people talking in the background (it happens), excessive bleed between microphones, or any sounds that you don't want to appear inadvertently in your mixes. You may have to solo tracks one at a time to find these unwanted sounds, but this is not really a waste of time (as long as there is not too much of this stuff going on) if you view it as a useful part of the process of familiarization that most people need to go through before getting creative with the mix (because you get to hear what is really happening on the tracks that you solo).

Noise gates can be used to reduce or remove bleed or other unwanted sounds in between the wanted sounds, but Pro Tools, like other DAWs, also makes it easy to cut out or mute regions where no music is playing. There are many ways to do this, starting with the most obvious, which would be to select the unwanted region and use the Cut command (Command-X on Mac, Control-X on Windows). If you want to remove everything between, say, a series of wood block or clave "hits," you can use the Separate Region at Transients command, available from the Edit Menu, which would be a lot faster than individually selecting and deleting the regions between each of these "hits"! Or you could insert a noise gate plug-in onto the tracks you want to clean up and use dynamics processing instead—which would be the analogue engineer's way of doing things.

> **Tip**
>
> Sometimes it is better to leave some, or even all, of the incidental noises in the recording: it can sound more natural or real that way and this can add to the appeal of the music. Recordings made with all the tracks tightly gated or with everything cut out of the tracks apart from the featured vocal and instrumental phrases can sound too "processed" for some people, me included. It's a bit like the difference between a wholesome homemade burger and the MacDuck variety.

Sorting the Tempos

It is possible that a client will bring you a session to mix that has no tempo set. Although you can just go ahead and mix without any reference to tempo or bars and beats, it is always a good idea to have a tempo for the music, or a tempo map if the tempo changes. This way, you can use Grid Mode whenever you would like to, and you can easily make edits at obvious places such as the beginning or the end of a particular bar or beat. So, it is worth spending some time with the Identify Beat command or using Beat Detective to create a tempo or tempo map before you start mixing.

Marking the Sections

Markers that say where the verse, chorus, middle, or whatever, sections start can be a invaluable timesaver on any mix session. If there are no markers to define the various sections of the music or song, then you should create a suitable set of markers at the outset of your mix session.

Grouping Tracks

You will need to group the tracks together for all sorts of reasons during your mix session. Maybe you want to change the fader levels of all the percussion tracks at the same time, for example. Or maybe you want to apply a delay-effect to all the backing vocals.

The most basic way of grouping tracks together—so that when you move one fader all the faders in the group move—is to use the basic Track Group feature where you select two or more tracks that you want to group in this way and press Command-G (Mac) or Control-G (Windows). The next step up from this is to create a subgroup by routing the outputs of a group of audio tracks to the input of an Auxiliary Input channel so that you can use that channel's fader to control the level of the subgroup. Both of these techniques have their uses and you will find many applications for these during your mixing sessions.

VCA Groups

If you are an experienced mixing engineer who is new to Pro Tools, you will probably know all about VCA Groups and will be pleased to know that these are available in Pro Tools. If not, you are probably wondering just what a VCA Group is. Briefly, VCA Groups were originally developed for analogue mixing consoles as a way of controlling a group of faders using a special control fader. Pro Tools VCA Master Faders emulate the behavior of these.

Conventional VCA Groups

To understand VCA Groups a little more thoroughly, you need to know something about how these were developed. Firstly, VCA is the abbreviation for a Voltage Controlled Amplifier. Mixing console designers realized that they could include a set of faders on an analogue console that could be used to send voltages to the amplifiers in selected audio channels so that one of these VCA Master Faders would control the audio levels for a group of mixer channels. A group of mixer channels being controlled in this way by a single VCA Master

Fader is then referred to as a VCA Group. No audio passes directly through the VCA Master Fader. VCA Master Fader is only used to set the level of the DC control voltage that is sent to control the levels of the faders in the group.

Using a single VCA Master Fader to control a number of mixer channel faders is much more efficient than creating a stereo subgroup using a stereo Aux channel with the individual mixer channel outputs routed to this. A VCA Master is much easier, and cheaper, to implement than an audio channel because it only has one fader that sets a voltage level and some routing and switching circuitry—it has no (relatively) expensive amplifier or signal processing circuitry.

When you lower the VCA Master Fader, you are effectively lowering the mixer channel faders in the group that you are controlling. This brings the advantage that any post-fader auxiliary Send feeds are lowered in turn, so the wet–dry balance of any effects that you are applying to these mixer channels stays the same. This is normally what you want to happen and is not the case when you have created a subgroup by routing the mixer channel outputs to an Auxiliary input channel, and you use that channel's fader to lower the level of the subgroup.

VCA Master Tracks in Pro Tools

(Pro Tools HD and Pro Tools with Complete Production Toolkit 2 Only)

As the Pro Tools reference guide explains, VCA Master tracks in Pro Tools emulate the operation of voltage-controlled amplifier channels on analogue consoles, where a VCA channel fader would be used to control, group, or offset the signal levels of other channels on the console. The way this works in Pro Tools is that you group a selection of tracks into a Mix group and then assign this group to a VCA Master track. This lets you control the output levels of all the VCA group's member tracks without the need to bus them to an Auxiliary Input track or to the same output path. You can create multiple, nested VCA groups and control the output levels of multiple sub-mixes at the same time—and you can conveniently automate a sub-mix by automating its VCA Master track. And a major advantage of using VCA Master tracks in Pro Tools is that they don't use up any of your precious mixer DSP resources.

> **Note**
>
> You can use a single VCA Master to control a selection of mixer channels that are routed to different outputs, which you cannot do by creating subgroups using Auxiliary Inputs.

The most obvious thing that a VCA fader assigned to a group of tracks allows you to do is to raise or lower the levels of all the individual tracks using just one fader. You can do this using an Auxiliary track, but using a VCA can be more

Figure 7.18
Sax Group
controlled by a
VCA.

appropriate. For example, a VCA allows you to group channels that are routed to different outputs, but Auxiliary tracks do not.

Using a VCA also has the advantage that changing the VCA fader level will not alter the balance between "wet" and "dry" for any tracks using post-fade Aux Sends, which would happen otherwise. This happens because when you reduce the fader levels, you are reducing the amount of "dry" signal in your mix—but you are not reducing the Send levels to the reverb—so the "wet" level stays the same.

You will also change the wet/dry balance if you insert a reverb plug-in on an Aux track to which a group of tracks is being bussed and then lower the Auxiliary track's fader, but not if you use a VCA fader to reduce the levels of all the individual faders in this group of tracks while leaving the Auxiliary track's fader untouched (without any gain change).

Because they don't pass audio, VCA Master tracks don't have inputs, outputs, inserts, or sends—they just have a fader along with Solo, Mute, Track

Record Enable, and Track Input Monitor buttons. They also have two popup selectors—one for Automation and one for Group Assignment—to let you select which Group of "slave" tracks to control. Note that on VCA Master tracks, the level meters indicate the highest level that has occurred on any of its individual tracks, not the summed level of all its slave tracks.

Keep in mind that unlike VCAs on traditional analogue consoles, VCA Master tracks in Pro Tools directly affect their slave tracks, so that the controls on each slave track always show their actual values. In other words, what you see on the slave tracks is what you've got! The levels of the faders show the actual levels as affected by the VCA Master. Even if a particular slave track is a member of more than one VCA-controlled group, the contribution of all the VCA Master Volume faders is summed on the slave track.

Because the slave tracks are grouped together first, using the standard Pro Tools Group command, you would expect that clicking an individual Mute or Solo would mute or solo all the Group tracks. However, when controlled by a VCA Master, these slave tracks operate individually, so clicking on just one slave track's Mute or Solo button will only affect that track.

> ## Note
>
> You can override this behavior and make these controls work as they normally would by deselecting the "Standard VCA Logic for Group Attributes" option in the Mixing section of the Pro Tools preferences window.

Also, the solo, mute, record-enable, and Track-Input status on each slave track will correspond with how these controls have been set on the VCA Master. So, for example, when you click Mute on the VCA Master, the individual Mute buttons become engaged on all the slave tracks, but in a partially "greyed-out" condition visually, and you will hear no sound. Slave tracks are said to be "implicitly-muted" when you mute the controlling VCA Master, because you are intending, or implying, that you want to mute these slave tracks. If any of the individual slave tracks were muted before you press the Mute on the VCA Master, these will remain muted and retain their normal visual appearance. To see how this looks, take a peek at the accompanying screenshot.

When you click Solo on the VCA Master this button will light up yellow and you will hear all its slave tracks in solo—but the individual slave track Solo buttons will not light up yellow. The way this works is that soloing a VCA Master will implicitly mute all tracks except its slave tracks, thereby indirectly soloing the slave tracks. Also, any explicit solos on the slave tracks will be cleared, leaving them indirectly soloed; and explicitly soloing a slave track while its VCA Master track is soloed will override the VCA Master solo.

Figure 7.19 Sax 2 was muted within the sax group. Then the VCA was muted. The screenshot shows the appearance of all the mute buttons. The mute button for Sax 3 has a different appearance to indicate that it is muted only as a consequence of muting the VCA Master and will not be muted when the VCA Master is unmuted.

Tip

The way that the Record enable and Track Input enable buttons work on the VCA Master is as follows: you can toggle the Record Enable and input monitor status on and off for any tracks that have been record-enabled using the VCA Master Record Enable or the VCA Master TrackInput button.

Using VCA's to Control Other VCA's

It can be very convenient to use VCA Master Faders to control two otherwise separate subgroups or VCA groups from one fader. For example, you might have a subgroup containing a brass section and another containing a string section. You may already be controlling these subgroups using two separate VCA Master Faders, or you may not. Whichever is the case, you can assign all the mixer channel faders to one single VCA Master that will let you raise or lower the brass and strings together.

Another example would be with a choir containing male voices and female voices. You would subgroup the male and female voices separately, set the balances between these, and then use a single VCA Master to raise or lower the level of the whole choir. Yet another example would be to create a subgroup containing drums, then to create a VCA Master to control the drums along with the bass guitar so that you could balance this rhythm section against the rest of the music.

Figure 7.20 A single Master VCA controls two individual VCA faders that in turn are controlling groups of saxes and guitars.

Edit Window Mix Automation

The Edit window lets you view automation and controller data either as a Track View to show one or other type of data, or in separate lanes under each track; in either case, you can view as many types of automation and controller data as you wish.

This lets you keep your preferred Track View (such as Region View) visible in the Edit window for each track, ready for last-minute edits while simultaneously viewing volume automation, for example, in a lane below each track. This can be very convenient during a mixing session when you want to make "tweaks" to the automation graphically onscreen in these Automation and Controller lanes instead of using the Mix window's real-time automation features.

Automation and Controller lanes can be shown or hidden under each track in the Edit window by clicking on the small arrow to the left of the Track controls in Medium or larger views.

In views smaller than Medium, you can show or hide Automation lanes using the Track options popup selector.

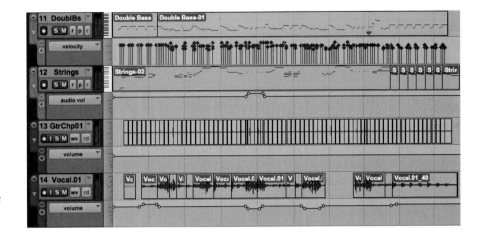

Figure 7.21
Automation and
Controller lanes
shown below the
tracks in the Edit
window.

Figure 7.22 Show or hide automation lanes using the arrow to the left of the Track controls.

Figure 7.23 A Track Options popup is located at the top left of each Track's controls area in the Edit window.

On audio tracks that are in Medium or larger view, you can use the popup Lane view selector in each lane to change the type of automation shown in the lane.

Figure 7.24
Lane view selector.

Choices include volume, volume trim, mute, and pan, along with automation controls for any sends or plug-ins that are active.

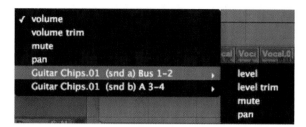

Figure 7.25
Lane view selector popup for an Audio track.

In the example given in the accompanying screenshot, an Instrument track is shown with a Controller lane open for MIDI Velocity and an Automation lane open for Audio Volume.

Figure 7.26 An Instrument track with Controller lanes for MIDI Velocity and Audio Volume. You can click on the "+" button to add another automation lane.

If you want to add more lanes below a track, you can click on the "+" button in an existing lane. To remove a lane click on the minus (–) button. The Lane view selector for an Instrument track lets you select from the various MIDI Controllers and Audio Automation types. The Lane view selector for a MIDI track lets you select from the various MIDI Controllers.

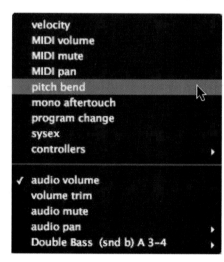

Figure 7.27
Lane view popup selector for an Instrument track shows the MIDI Controllers and audio automation types that can be selected.

To reorder lanes on-screen, drag the lane control sections to new positions in the Edit window. You can resize the heights of all the Automation and Controller lanes for a track by clicking and dragging the bottom line of any Lane Controls column up or down. To resize the height of a single Automation or Controller lane for a track, hold down the Control (Mac) or Start key (Windows) while you select the Lane Height setting. To increase or decrease the height of any lane that contains the Edit cursor or an Edit selection, hold down the Control (Mac) or Start key (Windows) and click the Up- or Down-Arrow key.

The Final Mix

Having completed all your preparations, the final balancing session often becomes relatively easy. In the early stages of your mixing session, you will have chosen most of the elements you want to work with and you will have set up the effects you want to use. It can take some time to set up sends to external effects units and Auxiliary tracks with chains of plug-ins to process the drums, the vocals, and the various instruments—especially if you are looking for that "special" reverb sound or combination of delays or if you are trying to find the right EQ settings and pan positions to make particular instruments "sit" properly in your mix, or stand out as features. With a typical pop song, you should allow at least half a day to get everything more or less in place. Then you might take a break for half an hour or so before coming back to do the final balancing of levels and tweaking of effects. Here, you should be concentrating on the most important tracks—typically the lead vocal and any featured solo instruments.

Make sure that all the tracks you are not using are either removed from the session or made inactive and hidden and set the rest to the minimum track height—just leaving the lead vocal, backing vocal, and perhaps some solo instruments at medium track heights. It is also a good idea to display the Volume Automation curves in the chosen tracks so that you can see what is going on and manually adjust the volumes at any stage during the mix session. Alternatively, you can use the Automation lanes to display volume or other automation or controller data in the Edit window.

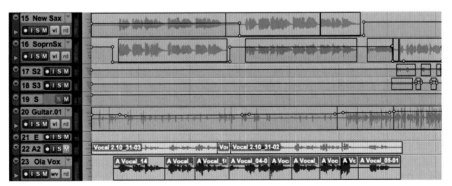

Figure 7.28
Pro Tools Edit window during a mixing session with the Volume automation curves for the Sax and Guitar visible—ready for the engineer to tweak.

If you are using the real-time automation features, you will start out in Auto Write mode. You probably won't get all your moves correct during the first pass, so you can use Auto Touch mode to refine the sections that need changing. As you get closer to what you are looking for, you might use the Trim features to finetune the settings even further. For example, if you like the way the guitar rhythm gets softer in the verse and louder in the choruses, but you decide that overall the guitar is too loud, you can use Trim mode to bring the overall level down on the guitar track while retaining the relative automation "moves."

I usually end up going into the Edit window to manually edit the automation breakpoints to achieve exactly the right result. Here, it is so easy to see exactly where the vocal and instrumental phrases lie and to edit the automation curves to do exactly what you want them to do.

> **Tip**
>
> To move all the automation breakpoints in a particular track or section of a track up or down while preserving relative levels, choose the Selector tool first and drag the mouse to highlight the range of interest. Then choose the Trimmer tool, point the mouse anywhere along the line of breakpoints, and drag the whole line upwards or downwards.

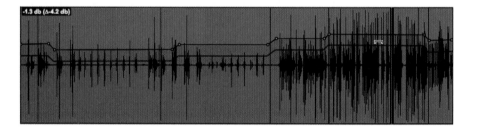

Figure 7.29
Manually trimming a selection of Automation breakpoints.

Once you have your mix sounding the way you want it, you will usually want to record this as a stereo "master" mix. You have various options here. You can record from the main stereo outputs of your Pro Tools interface into a stereo mastering recorder such as a CD-recorder or a DAT recorder, or even to a 1/2-inch or 1/4-inch analogue tape recorder.

Alternatively, you can use the Bounce to Disk command in Pro Tools to directly create a stereo file on disk. To do this, you mute everything but the tracks you want to bounce; make sure that all the levels, pans, and any effects and automation are in the way you want them to be; assign the outputs from all the tracks to the same pair of outputs; and then select the Bounce command from the File menu and bounce your mix from this output pair to disk. Although you can hear the bounce as it is being created, you can't adjust any controls during a "Bounce to Disk." So, you should only use this method if you don't need to interact with any mixer controls during the bounce. Also, you cannot include audio coming in from external inputs during a bounce to disk; if you need to do this, you will need to record to new audio tracks.

If you want to be able to adjust the mixer controls in real time while audio files are being written to disk or include audio coming in from external sources, you should record your mix to new tracks in Pro Tools. All you need to do is to route an additional output from your Master Fader to the input of an available stereo audio track in your Pro Tools session. Then mute the track onto which you are recording and listen to the mix in the normal way while you record your mix onto this track.

"To Dither or Not to Dither"—That Is the Question!

If you are mastering from a 24-bit session to a 24-bit digital recorder, or to analogue tape via 24-bit D/A converters, there is no need to apply dither. On the other hand, if you are mastering to a 16-bit medium—which can be a file on a disk or on an external recorder—you definitely should apply dither by inserting a dither plug-in on the Master Fader, which controls the stereo outputs that you are mixing or bouncing to.

Of course, if you are mixing to an analogue tape, there is no need to convert sample rates or bit-depths or to add dither. The D/A converters in the Pro Tools HD audio interfaces will convert the digital audio to analogue audio with very little loss of quality, no matter which sample rate or bit depth you are using, although you may prefer to use Apogee or PrismSound converters if you are looking for the very highest quality. Nevertheless, if you are mixing to an analogue tape, it is wise to simultaneously create a digital master so that you can make listening copies without generation loss. Of course, you won't hear the "audio signature" of the analogue master recording unless you make copies from this.

Mix Bus Compression

There has been a trend since the late 90s to add far too much compression and limiting on the mix bus so that the mixes sound as loud as possible when played back from a CD. Several multiband compression and limiting plug-ins are available for Pro Tools|HD that can be used to increase the average level by reducing peak levels and then raising the "make-up" gain. The danger here is that if the peak-to-average ratio is reduced too much, the music loses its "life" and sounds unpleasant. Yes, it sounds louder when played back from a CD—if you leave the volume control in the same position as for a "quieter" CD. But most people will make the volume the same by turning the knob to their preferred listening level, so all you will succeed in doing is to squash the "life" out of your music and run the risk of making the listener get tired of listening to it very quickly.

Mastering engineers will always prefer that you send final mixes without too much compression applied at the final mix stage—preferably with none on the mix bus. They use specialized equipment and high-quality monitoring systems, and they have much greater experience in this area, so they will almost always achieve much better results than you will be able to achieve yourself. And if you have already applied heavy compression or limiting to your mix, there is not much that the mastering engineer can do with this. Of course, if you like the sound of a particular mix bus compressor, you can always make one mix with the compressor and one mix without, so that the mastering engineer can choose whether to use this or not.

The Good, the Bad, and the Ugly

As examples of quality mixing and mastering techniques, just listen to the pristine audio quality with bags of dynamic range on Shelby Lynne's "Just A Little Lovin'" album or Dido's "Safe Trip Home" album. These albums will not fatigue your ear in any way, and will sound sweet for years to come!

If you listen to many of the today's commercial CD releases, even those re-issuing older material, you will notice that they sound extremely tiring, and tiresome, to listen to. A good (or should that be "bad"?) example would be Amy Winehouse's "Back to Black" album. Even though this has sold over 11 million copies, I still cannot listen to it for very long without experiencing listening fatigue because of the way the mixes have been compressed and limited during the production process.

A particularly ugly example would be the Tom Jones album "24 Hours." The main offender on this album seems to be the drum sounds, which just ruin the sound of all the mixes due to this ear-fatiguing effect, along with excessive limiting and compression used to make the average level of the music very high compared with the peak levels in the audio—thus destroying the natural dynamics in the sounds.

A Plea for Sanity

I hope that every Pro Tools user who reads this will join me in the fight against the loudness wars that are destroying the quality of many commercially released recordings today. It does not have to be this way! Practical advice from the experts is that engineers should mix and normalize to −3 dBFS; always avoid digital clipping; use low-level dynamics processing; use up-sampled limiting (or process in the analogue domain); and use up-sampled peak metering to make sure that the problems due to inter-sample peaks are avoided. It can also help to use loudness-calibrated monitoring and proper loudness metering. Using Bob Katz's K-system, or a return to the way music used to be mixed, or engineers aiming to have levels averaging around the 0 VU operating level, with up to 18 dB of headroom above this would allow music to sound much better than most of today's commercial releases. Above all, remember to use your ears!

Summary

So, here, we are at last, at the end of the music production process, hopefully listening back to wonderful music that we have recorded, edited, and mixed using Pro Tools. If you have read this book cover-to-cover, your head is probably full of things that you are trying to remember about how Pro Tools works. Don't worry: it will all fall into place for you eventually. Just expect that it will take some time. After all, "Rome wasn't built in a day," as they say in Venice.

To learn Pro Tools thoroughly, you need to work on as many different projects as you can, ideally with other people asking you to do things that you would not normally do yourself. Just like a musician can fall into the trap of practicing the same familiar licks and exercises over and over without learning anything new, a Pro Tools user can find that they are only working on one type of musical project for which they only need to use a restricted number of the features that Pro Tools has to offer.

As you start to build up your experience, take some time out to review various topics in this book. Learning keyboard commands and shortcuts is always very beneficial, and making sure that you are up to speed with all the different zooming and navigation methods is essential if you want to be able to work quickly.

Having a thorough understanding of how the meters and faders work, and how you can use Master Faders to prevent clipping on outputs and buses is probably the most important section in this chapter. After all, mixing is very much about getting the levels of the mix elements balanced to your satisfaction—without the digital waveform clipping that so often produces such harsh sounding mixes on commercially distributed music today. So I will leave you with these thoughts.

In This Chapter

8

New for Pro Tools 9

New Hardware and Software for Pro Tools 9 and HD 9

New Pro Tools HD Series interfaces, including the HD I/O, HD Omni, Sync HD, HD MADI, and Pre, were introduced in 2010 along with the Pro Tools|HD Native system. The Pro Tools|HD Native PCIe card enables you to run Pro Tools HD software with a lower-cost "native" solution using the host computer's processor rather than the dedicated DSP processors on Pro Tools|HD TDM systems. These are all described in some detail in Appendix 7, which can be found on the Website accompanying this book.

Pro Tools 9 and Pro Tools HD 9 software applications were also introduced in the final quarter of 2010.

Pro Tools|HD Native vs. Pro Tools|HD Hardware

Avid introduced Pro Tools|HD Native hardware systems shortly before the Pro Tools HD 9 software was released. Now that Pro Tools|HD Native has been introduced, you may be wondering why Pro Tools|HD systems are still available.

After all, a Pro Tools|HD Native Core system is significantly less expensive than the Pro Tools|HD Core system—less than half the price (at around £2510 + VAT compared with £5750 + VAT in the UK).

Well, without access to the DSP processors on the Pro Tools HD cards, you cannot run TDM plug-ins (which many professional users value very highly) and you will also lose the near-zero latency that these cards provide. So I expect that many professional users will continue to choose Pro Tools HD systems rather than the Pro Tools|HD Native systems.

The Avid Pro Tools|HD Native Core System supports up to four HD series interfaces (or one HD MADI interface) for up to 64 channels of I/O, and can playback up to 192/96/36 audio tracks at 48 kHz/96 kHz/192 kHz sample rates. It is also compatible with the original Digidesign "blue" HD interfaces (such as the 192 I/O and 96 I/O).

Pro Tools HD 1 systems provide just 32 channels of I/O and can playback 96/48/18 tracks at sample rates of 48/96/192 kHz. HD 2 gives you 64 channels of I/O and 192/96/36 tracks, while HD 3 gives you 96 channels of I/O and 192/96/36 tracks.

With Pro Tools|HD and HD|Native hardware, you can have up to 512 audio tracks in any session with up to 192 active voices (except with HD1, which allows only up to 96 voices). However, you have an additional ability to manually assign voices to tracks and use "voice borrowing." This allows tracks to share voices so that you can "voice" more than 192 tracks.

With PT8 and earlier versions, you could manually voice all 256 tracks in a session. In PT HD 9, you can only assign up to 336 voices to the 512 audio tracks that can be included in a session.

However, tracks that share a "voice" can playback, but not at the same time. The way this works is that the lowest numbered track takes priority and manually voiced tracks take priority over dynamically voiced tracks. This is really useful because as long as you don't have 192 things happening at once, you can use many more tracks and still hear everything playing back on these.

> ## Note
>
> The actual number of tracks that can be played back will also depend on the capabilities of the computer system, and particularly on the hard drive or hard drives in use. If the hard drives are not fast enough or if the number of tracks and edits are very high, the computer system and the hard drives may not be able to cope, and playback will be interrupted.

Pro Tools 9 vs. Pro Tools HD 9 Installation

There is just one unified Pro Tools 9 installer that will install the appropriate version of the software according to whether you have an iLok containing a Pro Tools HD 7 or 8 license attached to your computer or not. If you have that, Pro Tools HD 9 gets installed and if not, Pro Tools 9 software gets installed.

More specifically, if you install Pro Tools 9.0 on your Pro Tools|HD or Pro Tools|HD Native system, launching Pro Tools 9.0 will run the Pro Tools HD software (if a valid Pro Tools HD authorization is detected on a connected iLok and

Pro Tools|HD or Pro Tools|HD Native hardware is present). If you install Pro Tools on a system with any other hardware, launching Pro Tools 9.0 will run Pro Tools software (if a valid Pro Tools authorization is detected on a connected iLok).

- A Pro Tools HD 9.0 authorization lets you run Pro Tools HD 9.0 on a supported Mac or Windows computer with Pro Tools|HD or Pro Tools|HD Native hardware.
- A Pro Tools HD 9.0 authorization also authorizes Pro Tools 9.0 with Complete Production Toolkit 2 functionality on supported Mac or Windows systems without Pro Tools|HD or Pro Tools|HD Native hardware.
- A Pro Tools 9.0 authorization lets you run Pro Tools 9.0 on a supported Mac or Windows computer with a Pro Tools or M-Audio interface, or any third-party audio interface with supported Core Audio (Mac) or ASIO (Windows) device drivers (including the built-in audio on Mac computers using Core Audio).

Figure 8.1
Pro Tools 9 and
Pro Tools HD 9.

New Hardware System Definitions

Avid now defines Pro Tools HD systems to include Pro Tools HD (9.0) software with Pro Tools|HD or Pro Tools|HD Native hardware, while Pro Tools systems include Pro Tools (9.0) software with 003 or Digi 002 family audio interfaces, Eleven Rack, Mbox or M family audio interfaces, or M-Audio hardware.

There is also a third definition for those using third-party hardware: Pro Tools Core Audio/ASIO systems include Pro Tools 9.0 software with third-party audio interfaces with compatible Core Audio (Mac) or ASIO (Windows) drivers, including the built-in audio available on Mac computers (Core Audio).

Hence, Pro Tools 9 now lets you use any Apple Core Audio interface with a Mac or any Steinberg Audio Stream Input/Output (ASIO) interface with a Windows PC. Pro Tools can playback or record up to 32 channels of I/O with a Core Audio or ASIO interface; therefore, you are no longer restricted to using AVID Pro Tools interfaces with Pro Tools. If you intend to use an Avid Pro Tools interface, Core Audio and ASIO drivers are now available for most third-party software applications so that these applications will work fine with the Avid Pro Tools hardware.

For example, on my MacBook Pro with Digi 002 and Native Instruments Rig Kontrol hardware attached, both these options became available for use by the Pro Tools Audio Engine, together with the built-in audio hardware options and a new option—Pro Tools Aggregate I/O.

Figure 8.2
A part of the Pro Tools 9 Playback Engine dialog showing the choices available on my Macbook Pro with a Digi 002 and a Native Instruments Rig Kontrol attached.

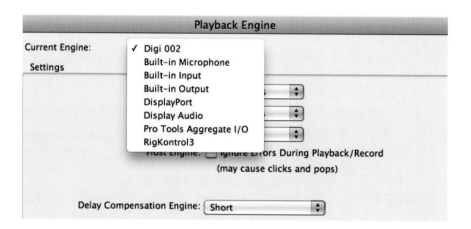

New Pro Tools System Settings

Pro Tools 9.0 lets you select the audio engine for the Pro Tools to use with your audio interface from the list of audio interfaces that are connected to your computer (assuming that these have compatible drivers installed). When you open the Playback Engine dialog from the Setup menu, you can select which "engine" you prefer, using the Current Engine popup selector at the top of the window.

Changing the Current Engine setting to use a different audio interface can be useful if you have multiple audio interfaces connected to your computer with different routing configurations in your studio, or if you want to prepare a session for use with a specific interface on a different system (for example you might want to prepare a session created on your HD system for use with the built-in audio on your Mac laptop).

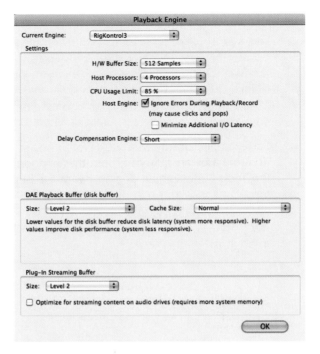

Figure 8.3
Playback Engine
for Pro Tools 9
system.

If you have a Pro Tools|HD or Pro Tools|HD Native system, the Current Engine defaults to using HD TDM or HD Native, so you would normally use this setting instead of the built-in audio. Similarly, if you have a Pro Tools or M-Audio audio interface such as the Mbox Pro or ProFire 2626, it makes sense to choose this type of interface. These use the Direct I/O engine for Pro Tools. And if you have a compatible CoreAudio (Mac) or ASIO (Windows) third-party audio interface, such as the Native Instruments Rig Kontrol, you would normally select this instead of the built-in audio.

If you have a compatible CoreAudio (Mac) or ASIO (Windows) third-party audio interface, such as the Native Instruments Rig Control, you would normally select this instead of any built-in audio.

Of course, if you want to use the built-in audio on your Mac, for example, you can select any of the available built-in options for playback on your particular computer; or select Pro Tools Aggregate I/O, which lets you use a combination of built-in inputs and outputs simultaneously for recording and monitoring.

Figure 8.4
Playback Engine
interface options.

Pro Tools Aggregate I/O

You can configure the Input and Output options for Pro Tools Aggregate I/O in the Mac Audio Setup, which can be accessed from the Pro Tools Hardware Setup. When you open the Hardware Setup dialog, just click on the Launch Setup App button to open the Audio Devices window of the Mac's Audio MIDI Setup application. When you select Pro Tools Aggregate I/O in this window, you are presented with a list of all the audio devices available to your Mac computer and you can tick the boxes in the window to enable your choices.

As Apple explains: "If you have several audio devices, you can use them as a single device known as an 'aggregate device.' By combining or 'aggregating' devices, you can increase audio capacity without purchasing more expensive multi-channeled audio equipment. For example, if you have an eight-channel audio device and a two-channel device, you can combine them to work as a single ten-channel device."

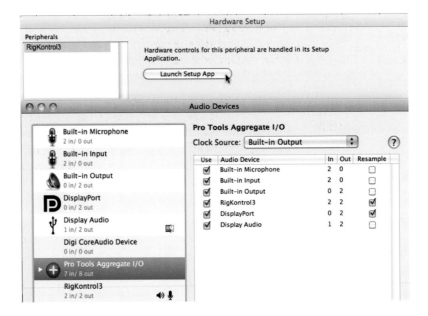

Figure 8.5
Pro Tools Hardware Setup dialog with Apple Audio Devices window in front.

Q: So, when should you use the Aggregate I/O and when should you use just one of the built-in audio options on the Mac?

A1: If you need simultaneous input for recording and output for playback and monitoring or if you want to combine devices to get more channels of I/O, then you should use the Pro Tools Aggregate I/O option.

A2: If you only need to playback audio for editing and mixing, choose an appropriate built-in audio output option, such as the Built-in Output option (the obvious one).

Complete Production Toolkit 2

Pro Tools 9 contains all the functionality that previously was only available in Pro Tools LE as paid Toolkit options (Music Production Toolkit 2, DV Toolkit 2, and Complete Production Toolkit), including all the advanced features of Beat Detective, DigiTranslator, and so on.

In addition, there is a new Toolkit option available for Pro Tools 9: Complete Production Toolkit 2. This adds most of the remaining high-end functions and features of Pro Tools HD 9 to Pro Tools 9 for users who do not have Pro Tools|HD systems.

> ### Note
> This toolkit is not intended for Pro Tools|HD or Pro Tools|HD Native system users, as Pro Tools HD software already includes the features in this package.

Complete Production Toolkit 2 costs £1430 + VAT in the UK which is a "handsome" price to pay. You need iLok authorizations for both Pro Tools 9 and Complete Production Toolkit 2 to be able to use the additional functions of the Toolkit, so you will have to pay for Pro Tools 9 if you don't have this already. However, for serious, professional users, there can be no doubt that it is worth paying the price.

Complete Production Toolkit 2 not only increases the total number of tracks you can have in a session to 512, and those that you can playback simultaneously to 192, it also adds support for up to 64 video tracks, gives you full VCA mixing capabilities, allows you to mix in up to 7.1 surround formats, and includes the Neyrinck SoundCode Stereo plug-in that allows surround sessions to be mixed down to stereo for 2-channel monitoring. So, using Pro Tools 9 with this Toolkit makes it possible to put a HD surround session with gazillions of tracks onto a laptop and tweak the surround mix while listening on stereo headphones on the beach—or even in the rainforest if this takes your fancy! Complete Production Toolkit 2 also includes the X-Form plug-in which offers the best time compression and expansion and the formant-corrected pitch-shifting available.

Pro Tools 9 with Complete Production Toolkit 2 will work with any Pro Tools audio interface, any M-Audio audio interface, or any third-party audio interface with supported Core Audio (Mac) or ASIO (Windows) device drivers.

> ### Note
> Registered owners of Complete Production Toolkit (version 1) are automatically eligible to use the features of Complete Production Toolkit 2 when they work with Pro Tools 9 software, so there's no need for them to upgrade. Avid also offers several options for existing Music Production Toolkit 2 and DV Toolkit 2 owners to upgrade to Complete Production Toolkit 2.

More Audio Tracks, More MIDI Tracks, and More Internal Mix Buses

Pro Tools 9 and Pro Tools HD 9 have both been upgraded to increase the number of tracks and buses available. It does take a bit of explaining to clarify how many of these are provided and how many of these can be used with the various different Pro Tools systems, so I have carefully extracted the facts and figures from Avid's website to present this information here for you.

First, let's look at the numbers of "voiceable" tracks available in Pro Tools 9 systems:

Pro Tools HD 9 and Pro Tools 9 with Complete Production Toolkit 2 can have up to 512 audio tracks in any session, 192 of which are "voiceable" while the rest are set to "off."

Pro Tools 9 sessions can have up to 128 audio tracks in a session but there are only 96 playback/record voices available in these systems, so only the first 96 (mono) or 48 (stereo), or some mixture thereof can be "voiced."

Note

You can have audio tracks beyond the system's "voiceable" track limit in a Pro Tools session, but these will be automatically set to Voice Off.

All versions of Pro Tools 9 provide up to 256 internal mix buses and up to 512 MIDI tracks.

The term "voices" in this context refers to the number of simultaneous playback and record tracks available with your system. Total "voiceable" tracks mean the maximum number of audio tracks that can share the available voices on your system. Keep in mind that mono tracks take up one voice while stereo and multichannel tracks take up one voice per channel.

According to Avid Pro Audio Application Specialist Simon Sherbourne, "The term "voiceable" comes from the Pro Tools HD (TDM) hardware, which allows you to manually assign voices to more than 192 tracks. There are still a total of 192 voices, but multiple tracks can share voices. With PT8 and earlier versions, you could manually voice all the 256 tracks in the session. With PT HD 9, you can assign up to 336 voices to the 512 audio tracks available in a session. Tracks that share a voice can playback, but the top one (in the edit window) has priority. When audio doesn't overlap, this effectively gives you more audio tracks that can playback."

Getting the tracks recorded all at the same time into your Pro Tools system is a different situation. And if you need to play the tracks back simultaneously through separate outputs, perhaps to feed these into an external mixing desk or into summing amplifiers, this is also a different situation. These capabilities depend on how many channels of input and output your Pro Tools system can support.

Always keep in mind that although the software may allow you to record or playback 128 or 96 voices, you can never input or output any more than, for example, 32 separate streams of audio with an HD 1 system at the same time—because the hardware only allows up to 32 I/O channels.

Avid's Simon Sherbourne comments, "That's correct in terms of hardware connections, as each Digilink port can handle up to 32 channels. The HD|Native card has two ports and so it can do 64 channels. Strictly speaking, you can input/record many more streams if you count ReWire."

Now let's look at the number of simultaneous tracks of playback and the numbers of I/O channels for HD systems.

The number of simultaneous tracks that can be played back using an HD1 system is 96 at 44.1 or 48 kHz sampling rate. This reduces to 48 tracks at 88.2/96 kHz and further reduces to 24 tracks at 176.4/192 kHz.

However, the maximum number of channels of input and output channels with Pro Tools|HD 1 systems is 32—so you cannot simultaneously record (or playback) more than 32 separate streams of audio into an HD 1 system.

Larger HD 2 systems provide 64 channels of I/O.

HD 3 systems provide 96 channels of I/O.

Expanded Pro Tools|HD systems provide even more; for example, HD 5 systems provide up to 160 channels of I/O.

These larger systems allow 192 simultaneous recording and playback voices at 44.1/48 kHz, 96 at 88.2/96, and 36 at 176.4/192.

In all cases, PT|HD systems can have up to 512 "voiceable" tracks.

So what does this mean in practice? It means that you could, for example, record 96 separate tracks in one pass, such as a large orchestra; then record 96 additional separate tracks in a second pass, perhaps a large choir using an HD3 system; and you would be able to playback all 192 of these tracks at the same time (although only through a maximum of 96 separate outputs).

If you have a very powerful computer with plenty of RAM and several fast hard drives, you could keep on recording and playing back additional tracks until you reach a maximum of 512—or until your system "falls over" and stop playing back any more because it could not keep up with the overheads of playing back such large numbers of tracks! So, there will eventually be a practical limit to your system's track playback capabilities, probably due to bandwidth limitations of your hard drives, which will eventually prevent playback. An important thing to know is that the Pro Tools software will not be a limiting factor until you try to exceed the limit of 512 tracks.

New Track and Send Output Selector Commands

Whenever I record several tracks of drums, percussion, guitars, vocals, or other groups of instruments, sooner or later I will almost certainly want to group these together and feed the outputs via a bus into an Auxiliary track where I can add delays or reverb to all the tracks in the group. This also allows me to control the overall level of the group in my mix using just one fader.

When you click on the Output selector for any track in Pro Tools 9, there are two new selections available: Hence, in addition to being able to assign the track output to any particular output or bus, you can, alternatively, assign the track output to the input of any existing track.

All the existing tracks in your session are listed in a drop-down sub-menu, accessible from a newly provided "track" menu item that allows you to choose from existing tracks.

> ## Tip
>
> Just one thing to watch out for here is the input of the destination track must be set to either an internal mix bus or to No Input in order to be available for the assignment in this way.

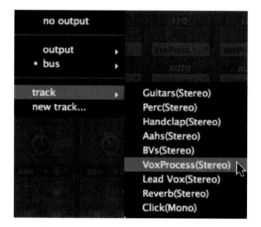

Figure 8.6 Assigning a track Output to the input of an existing track.

The second new selection, the "new track," lets you create a new Auxiliary Input, Audio, or Instrument track from the track or send Output selector.

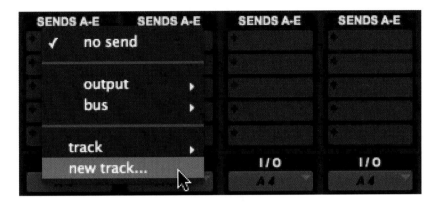

Figure 8.7 Creating a new track from a Send Output selector.

When you create a new track in this way, the source track (or send) output is automatically assigned to an available internal mix bus, as is the input for the newly created track—saving you the trouble of doing these things separately which you had to do with previous versions of this software.

If you need to set the track or send outputs (of the same send slot) for *all the tracks* in your Pro Tools session to the new track, simply Option-click (Mac) or Alt-click (Windows) as you use this "new track..." command.

More usefully, if you want to assign the outputs for *a selection of tracks*, such as all the vocal tracks, you can Option-Shift click (Mac) or Alt-Shift-click (Windows) to set all the selected tracks or send outputs (of the same send slot) to the new track.

When you choose the "new track..." command, a New Track dialog appears. You can use this to select the track width (mono, stereo, and surround), the track Type (Aux Input, Audio or Instrument Track), and the track Time Base (Samples or Ticks).

You can also name the track and tick or un-tick a box to choose whether to place this new track at the rightmost position in the Mix window or next to the current track. Note that if you have selected any track, so that the track name is highlighted, the new track will be positioned next to the right-most, currently selected track in the Mix window whether you tick this box or not.

Figure 8.8
Creating a new track from a Track output selector.

Figure 8.9
New Track dialog.

After you have chosen the track width, type, and time base and named the track, just click Create and the Pro Tools will create a new Auxiliary Input track named (in this example) "VoxProcess." It also creates an internal mix bus named "VoxProcess" (using the first available internal mix bus) and the output of the originating track or Send is automatically routed to the input of the new track using this mix bus.

This "new track…" command really speeds up the workflow when you are setting up a Pro Tools session; particularly, when you are preparing for a mix and need to quickly create Aux tracks to combine the outputs of many track pairs or groups for convenience.

Figure 8.10 A new Auxiliary track named "VoxProcess" has been created next to the rightmost selected track. Notice that the outputs of the selected tracks have all been set to the same bus that is then used as input to the Aux Track. This bus has automatically been given the same name, VoxProcess, which was typed into the New Track dialog while creating the Aux Track.

Assigning a Track Output to an Existing Auxiliary Input Track

There is one more convenient feature worth knowing about: when you want to assign a track output to an already-existing Auxiliary Input track, you can use the new "track" sub-menu to do this!

You might need to use this feature if, for example, you had already created an Auxiliary track at an earlier stage in your workflow, before you decided which tracks to assign to this.

Previously, you would have had to use the I/O Setup window to name a bus, then to have chosen this bus as the input to the Aux track and as the output from the track or tracks that you wished to assign to this.

Figure 8.11 An already existing Aux track without any input assigned.

In Pro Tools 9, you can simply choose this Auxiliary track from the existing tracks listed in the "track" sub-menu when you are assigning the output from a selected track or Send. Pro Tools will then automatically route the track or Send to the input of the existing Aux track using an available internal mix bus.

Figure 8.12 Routing a track output to an existing Auxiliary track using the "new track" command.

The great thing about using the "track" command to do this is that, as with the "new track..." command, Pro Tools chooses the next available internal mix bus and automatically names this mix bus using the name of the existing Auxiliary track that you have chosen.

System Settings vs. Session Settings

When exchanging a session between the Pro Tools systems, you will often need to reconfigure the session's I/O settings. So, it is worth considering where these settings are stored by different versions of the software.

Figure 8.13 Output from the track is routed to the input of the selected (existing) Auxiliary track and the mix bus is automatically named using the name of the existing Aux track.

In Pro Tools 8.1 and later versions (e.g., in Pro Tools 9), the I/O settings for the Input, Output, Insert, Mic Preamps, and H/W Insert Delay settings are regarded as the system settings and are always stored both with the system and with the session file—and can be recalled from either.

In Pro Tools 8.0.4 and lower versions, I/O settings are regarded as session settings and are stored and recalled using the session file. When you open a session that is created or edited on another system, any studio settings configured for your system will be overwritten by the settings stored with the newly opened session file.

The I/O Setup Dialog

When you open a session created or edited on another system, depending on what choice you have previously made in the I/O Setup dialog, you can determine whether or not the I/O settings stored with the session should overwrite the I/O settings stored with your system.

> **Note**
>
> The "Sessions overwrite current I/O Setup when opened" option is visible on the Input, Output Bus, and Insert pages of the I/O Setup dialog. Enabling or disabling this option in one page affects all other pages as well.

If you enable (tick the box for) the "Sessions overwrite current I/O Setup when opened" option and open a session, the Input, Output, and Insert I/O settings that are currently configured on your system will be overwritten by the I/O settings stored with that session. Specifically, the output bus paths of the session will remain mapped to the output paths saved with the session—so you may need to reconfigure these settings manually to match your current studio setup if the original session was set up using a different studio or an earlier configuration in your own studio.

This is the way that older (legacy) Pro Tools software worked, and this option is enabled by default.

When exchanging sessions with the systems that are running lower versions of Pro Tools, it is recommended that you should enable the "Sessions overwrite current I/O Setup when opened" option.

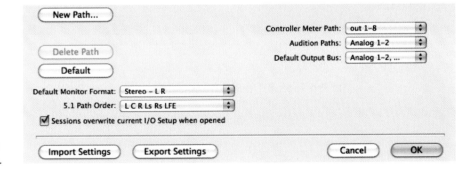

Figure 8.14
I/O Setup dialog showing Sessions Overwrite button.

If you disable (un-tick the box for) the "Sessions overwrite current I/O Setup when opened" option, Pro Tools recalls these settings from the system. In this case, the output bus paths of the session will automatically remap to the system output paths according to certain criteria (described in the reference manual).

When exchanging sessions among the systems that are running Pro Tools 8.1 or higher versions, it is generally recommended that you should disable the "Sessions overwrite current I/O Setup when opened" option. In this case, the system I/O settings stay the way you have set them up on your computer when you open a session from another system.

> ### Note
>
> When you open the sessions created in Pro Tools 8.0.4 (and lower) or in Pro Tools 8.1 (or higher), output paths from the legacy session are re-created as output buses. If the "Sessions overwrite current I/O Setup when opened" option is enabled, the output buses are mapped to session output paths. Obviously, if your hardware configuration has changed, you may need to manually reconfigure the output assignments using the I/O Setup grid.

New Output Buses in the I/O Setup

Pro Tools 9.0 provides two types of buses: up to 256 internal mix buses and a new, with version 9.0, output buses.

Typically, you will use internal mix buses to route audio signal from track outputs and Sends to other track inputs when you are setting up the effects sends and returns (e.g., bussing sends from the audio tracks to an Auxiliary Input track onto which you insert plug-in effects processing).

In addition to internal mix buses, Pro Tools 9.0 and higher versions provide output buses that appear along with the internal buses on the Bus page of the I/O Setup. When you create a new output path on the Outputs page, an output bus is automatically created and routed (mapped) to the corresponding output path. Output paths are assigned to the system's physical audio outputs in the I/O Setups Grid.

Because bus settings are saved with and recalled from the session, when moving sessions between Pro Tools systems, the output bus routing from the tracks and Sends in the Pro Tools mixer is maintained.

It is easy enough to see how everything works if you take a look at the I/O Setup dialog.

Select the tab for the Bus page and you will see a column on the left that lists the available output buses with their names displayed. These buses are listed in pairs with a revealing "arrow" at the left of each item that opens each bus pair to reveal the individual mono buses.

To the right of each listed bus pair, there is a checkbox that allows you to enable or disable this. The Format column indicates whether the bus is mono, stereo, or a surround format and a column is provided for text descriptions of these channels.

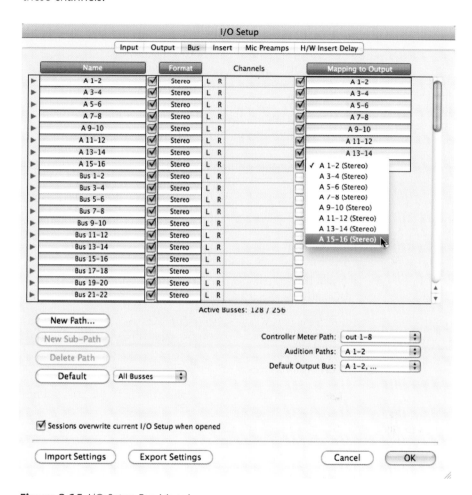

Figure 8.15 I/O Setup Bus Mappings.

At the far right, the Mapping to Output column provides popup selectors for each output pair to let you choose which physical output path to use with the output bus listed on the left. For example, I am using a Digi 002 interface that has 8 physical output pairs. By default, output bus pair 1-2 from Pro Tools is fed to the "outside world" using physical output pair 1-2 on the audio interface, output bus pair 3-4 feeds physical output pair 3-4, and so on. As can be seen in Fig. 8.15, a popup selector lets you choose which of the available physical outputs is mapped to which of the available output bus pairs from Pro Tools.

Importing I/O Settings

Two buttons at the bottom left of the I/O Setup dialog allow you to export or import the settings that you have made, storing these in a file on disk for future recall.

In earlier versions of Pro Tools, importing I/O Settings always imports settings for all pages in the I/O Setup.

In Pro Tools 9.0, the Import Settings button only imports the settings for the currently viewed page of the I/O Setup. For example, if you are viewing the Input page and use the import I/O Settings button, only the settings for the Input page are imported. This helps you to avoid overwriting any I/O Settings you have made on the other pages.

> **Tip**
>
> If you want to import settings to all the pages of I/O Setup in Pro Tools 9, simply Option-click (Mac) or Alt-click (Windows) on the Import Settings button.

If you have previously created a backup of your I/O settings by exporting your I/O settings, you can subsequently import these settings into any other session that you have opened.

When you import I/O Settings, you can choose to delete any unused path definitions before importing the new paths, or leave unused path definitions intact and add the new paths to the current I/O Setup configuration.

> **Note**
>
> You can also import I/O paths, path names, and other session data from a different session by using the Import Session Data command which can be accessed from the File menu.

Summary

The new features in Pro Tools 9 are there to make the workflow better and to make everything much easier for people who have to work with this software on a regular basis. The added features and capabilities make the upgrade very attractive for Pro Tools 8 users, and the fact that the Pro Tools software is no longer "tied" to Pro Tools hardware should open the door for a new generation of users to "come on board" and start using Pro Tools either instead of or alongside other popular DAW software.

For professional users with Pro Tools HD systems, the new features in version 9 make a lot of sense, especially for those who work on very large sessions and want to take copies of these out of the "big" studios to run them on a laptop using Pro Tools 9 with the Complete Production Toolkit 2. All-in-all, Pro Tools 9 looks like a winner to me!

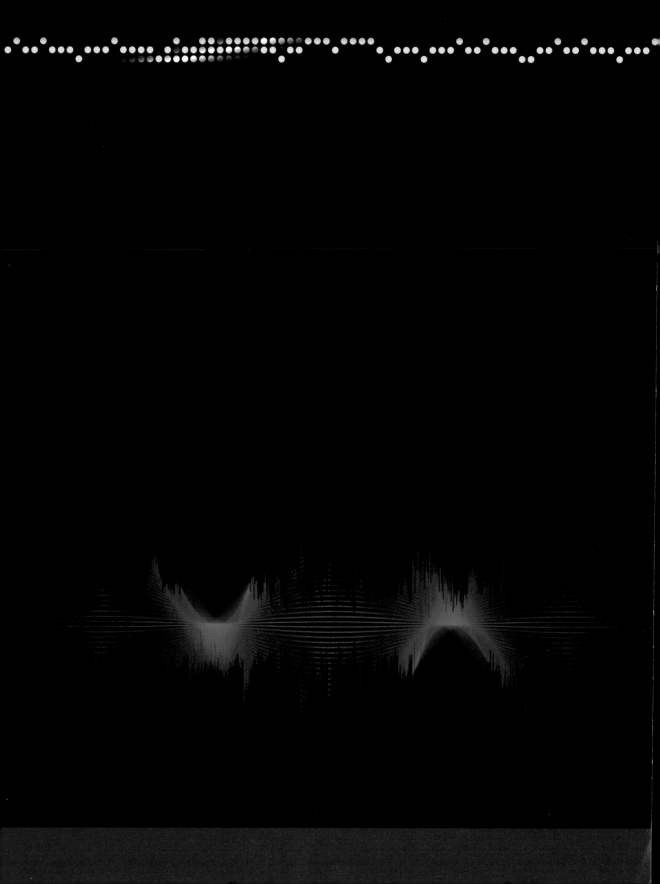

Index

Page numbers followed by *f* indicates a figure and *t* indicates a table.